T0305624

Fourier Analysis on Finite Groups with Applications in Signal Processing and System Design

Fourier Analysis on Finite Groups with Applications in Signal Processing and System Design

Radomir S. Stanković
Claudio Moraga
Jaakko T. Astola

IEEE PRESS

A JOHN WILEY & SONS, INC., PUBLICATION

Published by John Wiley & Sons, Inc., Hoboken, New Jersey.
Published simultaneously in Canada.

Limit of Liability/Disclaimer of Warranty: While the publisher and author have used their best efforts in preparing this book, they make no representations or warranties with respect to the accuracy or completeness of the contents of this book and specifically disclaim any implied warranties of merchantability or fitness for a particular purpose. No warranty may be created or extended by sales representatives or written sales materials. The advice and strategies contained herein may not be suitable for your situation. You should consult with a professional where appropriate. Neither the publisher nor author shall be liable for any loss of profit or any other commercial damages, including but not limited to special, incidental, consequential, or other damages.

For general information on our other products and services or for technical support, please contact our Customer Care Department within the U.S. at (800) 762-2974, outside the U.S. at (317) 572-3993 or fax (317) 572-4002.

Wiley also publishes its books in a variety of electronic formats. Some content that appears in print may not be available in electronic format. For information about Wiley products, visit our web site at www.wiley.com.

Library of Congress Cataloging-in-Publication is available.

ISBN-13 978-0-471-69463-2
ISBN-10 0-471-69463-0

Printed in the United States of America.

10 9 8 7 6 5 4 3 2 1

Preface

We believe that the group-theoretic approach to spectral techniques and, in particular, Fourier analysis, has many advantages, for instance, the possibility for a unified treatment of various seemingly unrelated classes of signals. This approach allows to extend the powerful methods of classical Fourier analysis to signals that are defined on very different algebraic structures that reflect the properties of the modelled phenomenon.

Spectral methods that are based on finite Abelian groups play a very important role in many applications in signal processing and logic design. In recent years the interest in developing methods that are based on Finite non-Abelian groups has been steadily growing, and already, there are many examples of cases where the spectral methods based only on Abelian groups do not provide the best performance.

This monograph reviews research by the authors in the area of abstract harmonic analysis on finite non-Abelian groups. Many of the results discussed have already appeared in somewhat different forms in journals and conference proceedings.

We have aimed for presenting the results here in a consistent and self-contained way, with a uniform notation and avoiding repetition of well-known results from abstract harmonic analysis, except when needed for derivation, discussion and appreciation of the results. However, the results are accompanied, where necessary or appropriate, with a short discussion including comments concerning their relationship to the existing results in the area.

The purpose of this monograph is to provide a basis for further study in abstract harmonic analysis on finite Abelian and non-Abelian groups and its applications.

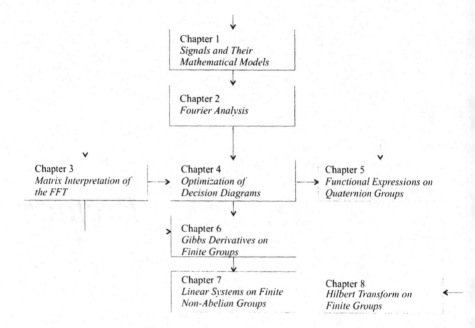

Fig. 0.1 Relationships among the chapters.

The monograph will hopefully stimulate new research that results in new methods and techniques to process signals modelled by functions on finite non-Abelian groups. Fig. 0.1 shows relationships among the chapters.

RADOMIR S. STANKOVIC, CLAUDIO MORAGA, JAAKKO T. ASTOLA

Niš, Dortmund, Tampere

Acknowledgments

Prof. Mark G. Karpovsky and Prof. Lazar A. Trachtenberg have traced in a series of publications chief directions in research in Fourier analysis on finite non-Abelian groups. We are following these directions in our research in the area, in particular in extending the theory of Gibbs differentiation to non-Abelian structures. For that, we are very indebted to them both.

The first author is very grateful to Prof. Paul L. Butzer, Dr. J. Edmund Gibbs, and Prof. Tsutomu Sasao for continuous support in studying and research work.

The authors thank Dragan Janković of Faculty of Electronics, University of Niš, Serbia, for programming and performing the experiments partially reported in this monograph.

A part of the work towards this monograph was done during the stay of R. S. Stanković at the Tampere International Center for Signal Processing (TICSP). The support and facilities provided by TICSP are gratefully acknowledged.

R.S.S., C.M, J.T.A.

Contents

List of Figures

List of Tables

Acronyms

ACDD	Arithmetic transform decision diagram
BDD	Binary decision diagram
BDT	Binary decision tree
DD	Decision diagram
DT	Decision tree
DFT	Discrete Fourier transform
FFT	Fast Fourier transform
FDD	Functional decision diagram
FNADD	Fourier decision diagram on finite non-Abelian groups
FNADT	Fourier decision tree on finite non-Abelian groups
FNAPDD	Fourier decision diagram on finite non-Abelian groups with preprocessing
FNAPDT	Fourier decision tree on finite non-Abelian groups with preprocessing
KDD	Kronecker decision diagram
mvMTDD	Matrix-valued multi-terminal decision diagram
MTBDD	Multi-terminal binary decision diagram
MTBDT	Multi-terminal binary decision tree
MDD	Multiple-place diagram
MTDD	Multi-terminal decision diagram
MTDT	Multi-terminal decision tree
nvMTDD	Number-valued multi-terminal decision diagram
PKDD	Pseudo-Kronecker decision diagram
QDD	Quaternary decision diagrams
SBDD	Shared binary decision diagrams
TVFG	Two-variable function generator
WDD	Walsh decision diagram

1

Signals and Their Mathematical Models

Humans interact with their environment using various physical processes. For example, the basic means of communication is through sound waves that are generated in the vocal tract and sensed by the ear. Visual communication is done by electromagnetic radiation that can be sensed by the eye. These as well as physical or mechanical interaction can be viewed as processes where a quantity; air pressure, electromagnetic field, physical bodies or their positions are changing as a function of time. These can in a natural way be interpreted as signals, mathematically described as continuous signals, functions of a real variable, often standing for the time. Natural phenomena, such as sound waves as the term indicate, often possess a periodic structure that can be described and analyzed using the powerful tools of Fourier analysis and other sophisticated concepts of mathematical analysis.

1.1 SYSTEMS

Many phenomena can be processed as continuous systems. A typical example is the room audio system, where the microphone picks up the changes in air pressure and converts it to electric voltage or current variations that are amplified and feed to loudspeakers to produce the same sound, but larger in volume. However, for long it has been known that under some restrictions a signal can be exactly reproduced from just knowing the values of the signal (function) at discrete but dense enough time instants. With the advent of digital computers this opened the way to digital processing of signal in which many of the limitations of physical electronic components can be avoided. This leads to the case where a (discrete) signal can be treated as a function on a finite cyclic group instead of the fields of real or complex numbers. The fast methods of computing different representations of discrete signals have enabled digital signal processing that has made possible the modern telecommunication systems and many other wonders of modern life.

Digital signal processing has leaped from its traditional area of processing digitally signals that used to be processed analogically to many new applications where the phenomena that are investigated can no more be represented as functions of a real or complex variable. The operations that are possible in the digital world are much more complicated than could be realized analogically. Also, the nature of signals may be very different from the original setting. A typical case is the investigation of logic functions using the same transforms as in digital signal processing and a more extreme example is the processing of the information coded in the DNA sequence using signal processing techniques.

When we apply the methods of Fourier analysis to a natural or man-made signals, the measurements or the data generated is represented as functions from a set to another. In principle, we could embed these sets in any mathematical structures, groups, rings, etc., for which the tools of Fourier analysis have been developed. However, to get full benefit from this powerful theory, the underlying structures should reflect at least some of the "true" properties of the signals, just as the cyclic group fits naturally to periodicity. Similarly, the dyadic group and the Walsh transform are able to capture properties of logic functions, and so useful in their representations. When more complex phenomena are studied it may not be possible to fully utilize the power of Fourier type methods if we restrict the domain of the signal to be an Abelian group and in certain fields where non-Abelian groups occur most naturally, such as crystallography, Fourier type methods based on non-Abelian groups are routinely used. In signal processing these methods are still not fully developed, but there are plenty of sporadic examples of the power of the theory.

In this book we concentrate in presenting (the theory of) Fourier methods over non-Abelian groups for signal processing and logic design. However, we believe that in due time there will be many more applications in the vast range of topics in which signal processing methods are applied.

1.2 SIGNALS

When we observe physical signals, changes in air pressure, electromagnetic field, etc., in analog form using some recording device, the recorded signal is only an approximation of the original due to the errors inherent in any sensors. Likewise, even if we assume that the original physical signal satisfied the requirements of the sampling theorem for exact reproduction, our sampling devices have their inherent errors and, thus, only an approximation of the digital equivalent of the original physical signal can be captured. However, as long as the errors are smaller than the accuracy required for extracting the relevant information the system is fine for practical purposes, and a key element in engineering practice is to balance the cost and performance of the overall system.

1.3 MATHEMATICAL MODELS OF SIGNALS

Since the signals are physical processes which spread in space-time they are best modelled by elements of some function spaces. To keep the connection to the real world we usually model the inaccuracies as random quantities (noise) following some probability distribution. Often it is necessary to view the "noisless" signal also as a random process that has a joint distribution with the noise process. There is an extensive literature on fundamentals of stochastic signals and the problems that are related to sampling and estimation of signals [27], [39]. Nevertheless the essence of what is now called the sampling theorem was also known to the earlier mathematicians. The reader is referred to [4], [16], [20], [50] for some discussions about the history, different formulations, extensions and generalizations of the sampling theorem. For the sampling theorem on the dyadic group see [35] and later [21]. The extension of the theory to arbitrary locally compact Abelian group is given in [26]. An interpretation of the sampling theorem in Fourier analysis on finite dyadic group is given in [31] and extended to arbitrary finite Abelian and non-Abelian groups in [47] and [50], respectively.

In engineering practice the signals are modelled by complex functions of real variables and usually called continuous signals. Those represented by discrete functions, i.e., by functions whose variables are taken from discrete sets, are called discrete signals. These also are divided into two subclasses depending on the range of values the signals can take.

The continuous signals of a real amplitude are analog signals, while the discrete signals whose amplitudes belong to some finite sets are digital signals.

To take advantage of similar powerful mathematical machinery in dealing with discrete signals, it is necessary to impose some algebraic structure on their domain as well as range. In this setting the signals are defined as functions on groups into fields. Moreover, as it has been shown in [52], the structure of a group is the weakest structure on the domain of a signal that still provides a practically tractable model for most of the signal processing and system theory tasks.

We consider discrete signals that are defined on some discrete groups, usually identified with the group of integers Z, or with some group Z_p of integers modulo p. In other words, the discrete signals are as functions $f : Z \rightarrow X$, or $f : Z_p \rightarrow Z_p$, where X may be the field of complex numbers C, the field of real numbers R, or the group of integers Z, or some finite field. For example, among Abelian groups, the dyadic group and finite dyadic groups G_{2^n}, $n \in N$, have attained a lot of interest, see for example [1], [3], since the Walsh functions [55], the group characters of these groups [12], and their discrete counterparts, the discrete Walsh functions, take two values $+1$ and -1 and, therefore, the calculation of the Walsh-Fourier spectra can be carried out without multiplication.

However, there are real-life signals and systems which are more naturally modelled by functions and, respectively, relations between functions on non-Abelian groups. We will mention some related with electrical engineering practice. Some other examples of such problems are discussed in [5].

As is noted in [23], there are examples in pattern recognition for binary images, which may be considered as a problem of realization of binary matrices, in synthesis of rearrangeable switching networks whose outputs depend on the permutation of input terminals [15], [34], in interconnecting telephone lines, etc. An application of non-Abelian groups in linear systems theory is in the approximation of a linear time-invariant system by a system whose input and output are functions defined on non-Abelian groups [24]. See, also [41], [42], [43], [44], [52].

The application of non-Abelian groups in filtering is discussed in [25], where a general model of a suboptimal Wiener filter over a group is defined. It is shown that, with respect to some criteria, the use of non-Abelian groups may be more advantageous that the use of an Abelian group. For example, in some cases the use of the Fourier transform on various non-Abelian groups results in improving statistical performance of the filter as compared to the DFT. See, also [51].

The fast Fourier transform on finite non-Abelian groups [23], [38], has been widely used in different applications [24], [25]. It may be said that for finite non-Abelian groups the quaternion group has a role equal to that played by the finite dyadic group among Abelian groups [50]. Similarly as with the Walsh transform, i.e., the Fourier transform on finite dyadic groups, the calculation of the Fourier transform on the quaternion does not require the multiplication. Regarding the efficiency of the fast Fourier transform on groups, it has been shown [38] for sample evaluations with different groups that in a multiprocessor environment the use of non-Abelian groups, for example quaternions, may result in many cases in optimal, fastest, performance of the FFT. Moreover, as is shown in [38], the quaternion groups as components of the direct product for the domain group G, in many cases, exhibit optimal performance in the accuracy of calculation.

These performances have been estimated taking into consideration the number of arithmetic operations, the number of interprocessor data transfers, and the number of communication lines operating in parallel. In this setting, looking for a suitable finite group structure G which should be imposed on the domain of a discrete signal, it has been shown that the combination of small cyclic groups C_2 and quaternion groups in the direct product for G results in groups exhibiting, in most cases, the fastest algorithms for the computation of the Fourier transform.

In practical applications, we often refer to topological properties of the algebraic structures we use for mathematical models of signals and systems. The space-time topology of the produced solutions stems from the topology (in the mathematical sense) of the related algebraic structures. It is intersting [14] that some important mathematical notions have been introduced first on more complicated structures, and then extended or transferred to the simpler cases. Differential operators could be mentioned as an example. Concept of Newton-Leibniz derivative, the notion was introduced first for real functions, although the continuum of the real line R is one of the most sophisticated algebraic structures (though the richness of the structure was not fully appreciated at that time). Extension of differentiation to the simple case of finite dyadic groups, was done about two centuries after the first vague ideas of differentiators and their applications in estimating the rate of change and the direction of change of a signal [13]. Moreover, it was motivated by the requirements of technology related to the interest in various applications of two-valued discrete Walsh functions in transmitting and processing binary coded signals and their realizations within prevalent two-stable state circuits environment. The support set of finite dyadic groups, the n-th order direct product of the basic cyclic group of order 2, produces a binary coding of the sequence of first non-negative integers less than 2^n, representing a base for the Boolean topologies often used in system design, including the logic design as a particular example of systems devoted the processing of a special class of signals, the logic signals [18], [19]. The restriction of the order to 2^n, and some other inconveniences of the Boolean topology, motivated the recent interest in topologies derived from the binary coding of Fibonacci sequences and their applications [1], [8], [9], [10], [17],[40]. Use of these structures permits introduction of new Fourier-like transforms [1], [7], [11], enriching the class of transforms appearing in Nature and computers [54], [56], [57]. Various extensions and generalizations of the representations of discrete signals and spectral methods in terms of different systems of not necessarily orthogonal basic functions on Abelian groups [2], and the use of non-Abelian groups in signal processing and related areas, suggested probably first in the introduction of [22], offer new interesting research topics as is shown, for example, in [6], [28], [29], [30], [32], [33], [36]. For these reasons, we have found it interesting to study Fourier transforms on finite non-Abelian groups, and Fourier-like or generalized discrete Fourier transforms [24] on the direct product of finite not necessarily Abelian groups [45], [46], [48], [49]. Some recent results in this area are discussed in [37].

These transforms are defined in terms of basic functions generated as the Kronecker product of unitary irreducible representations of subgroups in the domain groups. This way of generalizing the Fourier transform ensures the existence of fast algorithms for efficient calculation of spectral coefficients in terms of space and time. We call all these transforms Fourier transforms, with the excuse that efficient computation is, in many applications, stronger requirement, than possessing counterparts of all the deep properties of the Fourier transform on R. We also consider the Gibbs derivatives on finite non-Abelian groups, since they extend the notion of differentiation to functions on finite groups through a generalization of the relationship between the Newton-Leibniz derivative and Fourier coefficients in Fourier analysis on R.

REFERENCES

1. Agaian, S., Astola, J., Egiazarian, K., *Binary Polynomial Transforms and Non-linear Digital Filters*, Marcel Dekker, New York 1995.

2. Aizenberg, N.N., Aizenberg, I.N., Egiazarian, K., Astola. J., "An introduction to algebraic theory of discrete signals", *Proc. First Int. Workshop on Transforms and Filter Banks*, TICSP Series, No. 1, Tampere, February 23-25, 1998, 70-94.

3. Beauchamp, K.G., *Applications of Walsh and Related Functions: With an Introduction to Sequency Theory*, Academic Press, New York, 1984.

4. Butzer, P.L., Splettstösser, W., Stens, R.L., "The sampling theorem and linear prediction in signal analysis", *Jber. d. Dt. Math. -Verein.*, 90, 1988, 1-70.

5. Chirikjian, G.S., Kyatkin, A.B., *Engineering Applications of Noncommutative Harmonic Analysis*, CRC Press, Boca Raton, FL, 2000.

6. Creutzburg, R., Labunets, V.G., Labunets, E.V., "Algebraic foundations of an abstract harmonic analysis of signals and systems", *Proc. First Int. Workshop on Transforms and Filter Banks*, TICSP Series, No. 1, Tampere, February 23-25, 1998, 30-69.

7. Egiazarian, K., Astola, J., "Discrete orthogonal transforms bassed on Fibonacci-type recursion", *Proc. IEEE Digital Signal Processing Workshop (DSPWS-96)*, Norway, 1996.

8. Egiazarian, K., Astola, J., "Generalized Fibonacci cubes and trees for DSP applications", *Proc. ISCAS-96*, Atlanta, Georgia, May 1996.

9. Egiazarian. K., Astola, J., "On generalized Fibonacci cubes and unitary transforms", *Applicable Algebra in Engineering, Communication and Computing*, Vol. AAECC 8, 1997, 371-377.

10. Egiazarian, K., Gevorkian, D., Astola, J., "Time-varying filter banks and multiresolution transforms based on generalized Fibonacci topology", *Proc. 5th IEEE Int. Workshop on Intelligent Signal Proc. and Communication Systems*, Kuala Lumpur, Malaysia, 11-13 Nov. 1997, S16.5.1-S16.5.4.

11. Egiazarian, K., Astola. J., Agaian, S., "Orthogonal transforms based on generalized Fibonacci recursions", *Proc. 2nd Int. Workshop on Spectral Techniques and Filter Banks*, Brandenburg, Germany, March 5-7, 1999.

12. Fine, N.J., "On the Walsh functions", *Trans. Amer. Math. Soc.*, Vol. 65, No. 3, 1949, 372-414.

13. Gibbs, J.E., "Walsh spectrometry, a form of spectral analysis well suited to binary digital computation", *NPL DES Rept.*, Teddington, Middlesex, United Kingdom, July 1967.

14. Gibbs, J.E., Simpson, J., "Differentiation on finite Abelian groups", *National Physical Lab.*, Teddington, England, DES Rept, No.14, 1974.

15. Harada, K., "Sequental permutation networks", *IEEE Trans. Computers*, Vol.C-21, 472-479, 1972.

16. Higgins, J.R., "Five short stories about the cardinal series", *Bull Amer. Math. Soc.*, 12, 1985, 45-89.

17. Hsu, W.-J., "Fibonacci cubes-a new interconnection topology", *IEEE Trans. Parallel Distrib. Syst.*, Vol. 4, 1993, 3-12.

18. Hurst, S.L., *The Logical Processing of Digital Signals*, Crane Russak and Edvard Arnold, Basel and Bristol, 1978.

19. Hurst, S.L., Miller, D.M., Muzio, J.C., *Spectral Techniques in Digital Logic*, Academic Press, New York, 1985.

20. Jerri, A.J., "The Shannon sampling theorem-its various extensions and applications: a tutorial review", *Proc. IEEE*, Vol. 65, No. 11, 1977, 1565-1596.

21. Kak, S.C., "Sampling theorem in Walsh-Fourier analysis", *Electron. Lett.*, Vol. 6, July 1970, 447-448.

22. Karpovsky, M.G., *Finite Orthogonal Series in the Design of Digital Devices*, John Wiley and Sons and JUP, New York and Jerusalem, 1976.

23. Karpovsky, M.G., "Fast Fourier transforms on finite non-Abelian groups", *IEEE Trans. Computers*, Vol. C-26, 1028-1030, 1977.

24. Karpovsky, M.G., Trachtenberg, E.A., "Some optimization problems for convolution systems over finite groups", *Inform. and Control*, 34, 3, 227-247, 1977.

25. Karpovsky, M.G., Trachtenberg, E.A., "Statistical and computational performance of a class of generalized Wiener filters", *IEEE Trans. Inform. Theory*, Vol. IT-32, 1986.

26. Kluvànek, I., "Sampling theorem in abstract harmonic analysis", *Mat,-Fiz. Časopis*, Sloven. Akad. Vied., 15, 1965, 43-48.

27. Kotel'nikov, V.A., "On the carrying capacity of the 'ether' and wire in telecommunications", *Material for the First All-Union Conference on Questions of Communication*, Izd. Red. Upr. Svazy RKKA, Moscow, 1933 (in Russian).

28. Kyatkin, A.B., Chirikjian, G.S., "Algorithms for fast convolutions on motion groups", *App. Comp. Harm. Anal.*, 9, 220-241, 2000.

29. Labunets, V.G., *Algebraic Theory of Signals and Systems-Computer Signal Processing*, Ural State University Press, Sverdlovsk, 1984 (in Russian).

30. Labunets, V.G., "Fast Fourier transform on generalized dihedral groups", in *Design Automation Theory and Methods*, Institute of Technical Cybernetics of the Belorussian Academy of Sciences Press, Minsk, 1985, 46-58 (in Russian).

31. Le Dinh, C.T., Le, P., Goulet, R., "Sampling expansions in discrete and finite Walsh-Fourier analysis", *Proc. 1972 Symp. Applic. Walsh Functions*, Washington, DC, 1972, 265-271.

32. Maslen, D.K., "Efficient computation of Fourier transforms on compact groups", *J. Fourier Anal. Appl.*, 1998, 19-52.

33. Moore, C., Rockmore, D.N., Russell, A., *Generic Quantum Fourier Transforms*, Tech. Rept. quant-ph/0304064, Quantum Physics e-print Archive, 2003.

34. Opferman, D.C., Tsao, Wu, N.T., "On a class of rearrangeable switching networks", *Bell System Tech. J.*, 5C, 1579-1618, 1971.

35. Pichler, F.R., "Sampling theorem with respect to Walsh-Fourier analysis", Appendix B in Reports Walsh Functions and Linear System Theory, Elect. Eng. Dept., Univ. of Maryland, College Park, MD, May 1970.

36. Rockmore, D.N., "Some applications of generalized FFTs", An appendix w/D. Healy, in Finkelstein, L., Kantor, W., (eds.), *Proc. of the DIMACS Workshop on Groups and Computation*, June 7-10, 1995, published 1997, 329-369.

37. Rockmore, D.N., "Recent progress in applications in group FFTs", in J. Byrnes, G. Ostheimer, (eds.), *Computational Noncomutative Algebra and Applications*, NATO Science Series: Mathematics, Physics and Chemistry, Springer, Vol. 136, 2004.

38. Roziner, T.D., Karpovsky, M.G., Trachtenberg, L.A., "Fast Fourier transform over finite groups by multiprocessor systems", *IEEE Trans. Acoust., Speech, Signal Processing*, Vol. ASSP-38, No. 2, 226-240, 1990.

39. Shannon, C.E., "A mathematical theory of communication", *Bell Systems Tech. J.*, Vol. 27, 1948, 379-423.

40. Stakhov, A.P., *Algorithmic Measurement Theory*, Znanie, Moscow, No. 6, 1979, 64 p. (in Russian).

41. Stanković, R.S., "Linear harmonic translation invariant systems on finite non-Abelian groups", R. Trappl, Ed., *Cybernetics and Systems Research*, North-Holland, Amsterdam, 1986.

42. Stanković, R.S., "A note on differential operators on finite non-Abelian groups," *Applicable Anal.*, 21, 1986, 31-41.

43. Stanković, R.S., "A note on spectral theory on finite non-Abelian groups," *3rd Int. Workshop on Spectral Techniques*, Univ. Dortmund, 1988, Oct. 4-6, 1988.

44. Stanković, R.S., "Fast algorithms for calculation of Gibbs derivatives on finite groups," *Approx. Theory and Its Applications*, 7, 2, June 1991, 1-19.

45. Stanković, R.S., Astola, J.T., "Design of decision diagrams with increased functionality of nodes through group theory", *IEICE Trans. Fundamentals*, Vol. E86-A, No. 3, 2003, 693-703.

46. Stanković, R.S., Astola, J.T., *Spectral Interpretation of Decision Diagrams*, Springer, New York, 2003.

47. Stanković, R.S., Stanković, M.S., "Sampling expansions for complex-valued functions on finite Abelian groups", *Automatika*, 25, 3-4, 147-150, 1984.

48. Stanković, R.S., Milenović, D., Janković, D., "Quaternion groups versus dyadic groups in representations and processing of switching fucntions", *Proc. 20th Int. Symp. on Multiple-Valued Logic*, Freiburg im Breisgau, Germany, May 20-22, 1999, 19-23.

49. Stanković, R.S., Moraga, C., Astola, J.T., "From Fourier expansions to arithmetic-Haar expressions on quaternion groups", *Applicable Algebra in Engineering, Communication and Computing*, Vol. AAECC 12, 2001, 227-253.

50. Stanković, R.S., Stojić, M.R., Bogdanović, M.R., *Fourier Representation of Signals*, Naučna knjiga, Beograd, 1988, (in Serbian).

51. Trachtenberg, E.A., "Systems over finite groups as suboptimal filters: a comparative study", P.A. Fuhrmann, Ed., *Proc. 5th Int. Symp. Math. Theory of Systems and Networks*, Springer-Verlag, Beer-Sheva, Israel, 1983, 856-863.

52. Trachtenberg, E.A., "SVD of Frobenius matrices for approximate and multiobjective signal processing tasks", in E.F. Deprettere, Ed., *SVD and Signal Processing*, Elsevier North-Holland, Amsterdam/New York, 1988, 331-345.

53. Trachtenberg, E.A., "Applications of Fourier Analysis on Groups in Engineering Practices", in Stanković, R.S., Stojić, M.R., Stanković, M.S., (eds.), *Recent Developments in Abstract Harmonic Analysis with Applications in Signal Processing*, Nauka, Belgrade and Elektronski fakultet, Niš, 1996, 331-403.

54. Trakhtman, A.M., Trakhtman, V.A., *Fundamentals of the Theory of Discrete Signals Defined on Finite Intervals*, Sov. Radio, Moscow, 1975 (in Russian).

55. Walsh, J.L., "A closed set of orthogonal functions", *Amer. J. Math.*, 45, 1923, 5-24.

56. Yaroslavsky, L., *Digital Picture Processing, an Introduction*, Springer Verlag, Heidelberg, 1985.

57. Yaroslavsky, L., "Transforms in Nature and computers: origin, discrete representation, synthesis and fast algorithms", *Proc. First Int. Workshop on Transforms and Filter Banks*, TICSP Series, No. 1, Tampere, February 23-25, 1998, 3-29.

2

Fourier Analysis on Non-Abelian Groups

In this chapter, we present a brief introduction to the group representation theory and harmonic analysis on finite not necessarily Abelian groups. For more details the reader is referred to the voluminous literature on abstract harmonic analysis, e.g. [2], [6], [8], [17], [18].

The main idea of abstract harmonic analysis is to decompose a complicated function f into pieces that reflect the structure of the group G on which f is defined. The goal is to make some difficult analysis manageable [11].

The most widely used groups probably are the real line R and the circle $R/2\pi Z$, where Z is the set of integers.

In the case of the group $G = R/2\pi Z$, a given function f on $(-\pi, \pi)$ is decomposed as

$$f(x) = \sum_{n=-\infty}^{\infty} c_n e^{inx},$$

$$c_n = \frac{1}{2\pi} \int_{-\pi}^{\pi} f(x) e^{-inx} dx$$

and obviously the same representation extends to the periodic extensions of f.

For aperiodic functions on the real line R we have the direct and inverse Fourier transforms

$$S_f(w) = \int_{-\infty}^{\infty} f(x) e^{-2\pi iwx} dx,$$

$$f(x) = \int_{-\infty}^{\infty} S_f(w) e^{2\pi iwx} dw.$$

From the mathematical topology point of view, the real line R is a locally compact Abelian group and the theory of Fourier analysis can been extended to such groups if the exponential functions used in Fourier analysis on R are replaced by the group representations. Compact groups form a subset of the set of locally compact groups. In the following, we briefly introduce the basic concepts of group representations and Fourier analysis on finite groups, and then discuss in more details Fourier analysis on finite non-Abelian groups. We present the basic definitions and and properties as well as proofs of these properties. The purpose is to give the reader some insight to the representation theory in order to clarify the properties of the structures that we use later.

2.1 REPRESENTATIONS OF GROUPS

A representation of a group G on a complex vector space V is a correspondence between the abstract group G and a subgroup of the "concrete" group of linear transformations of V, that is, representation is a homomorphism of G into the group of invertible linear transformations on V. Often the group G and the space V are topologized and the group actions are then assumed to be continuous.

In the case of finite groups, the linear transformations are usually identified with matrices. In this setting the following definition of group representations can be introduced.

The general linear group $GL(n, P)$ is the group of $(n \times n)$ invertible matrices (n is a natural number) with respect to matrix multiplication, with entries in a field P that can be the field of complex numbers C or a finite field F_q where q is power of a prime p. Thus,

$$GL(n, P) = \{\mathbf{A} \in P_{n \times n} | \det(\mathbf{A}) = |\mathbf{A}| \neq 0\}.$$

Definition 2.1 *(Group representations)*
A finite dimensional representation of a finite group G is a group homomorphism $\mathbf{R} : G \to GL(n, C)$.

Notice that for a given $x \in G$, $\mathbf{R}(x)$ stands for an $(n \times n)$ matrix $\mathbf{R}(x) = [R_{i,j}]$, $i, j = 1, \ldots, n$. The matrix entries $R_{i,j}$ of $\mathbf{R}(x)$ are continuous functions in discrete topology, analogous to trigonometric functions on the circle or the exponential functions $\exp(2\pi i/n)$ in terms of which the classical Fourier analysis has been defined. Therefore, they will be used to define the Fourier transform on G.

Because any finite group of order N is isomorphic to a subgroup of the symmetric group S_N, the group of permutations of N objects, the elements of which can be explicitly listed as N (unitary) $(N \times N)$ permutation matrices, there are always nontrivial representations.

Every finite-dimensional representation is equivalent (similar) to a representation by unitary matrices [8]. Thus, if G is a finite group, every representation is equivalent to a unitary representation. Recall that unitary matrices preserve the inner product defined for two vectors x and y in C^n in a usual manner as $\langle x, y \rangle = \overline{x}^T y$, where $\overline{x}$

is the complex-conjugate of x and T denotes the transposition. Thus, for a unitary matrix $\mathbf{Q}$ it holds $\langle \mathbf{Q}x, \mathbf{Q}y \rangle = \langle x, y \rangle$ for all $x, y \in C^n$. We denote by $U(n)$ the multiplicative group of $(n \times n)$ unitary matrices, i.e.,

$$U(n) = \{\mathbf{Q} \in GL(n, C) | \overline{\mathbf{Q}}^T \mathbf{Q} = \mathbf{I}\},$$

where $\mathbf{I}$ is the identity matrix.

Let $\mathbf{R}$ and $\mathbf{S}$ be representations of degree n. If there is a subspace W of C^n such that $SW \subseteq W$ and $\mathbf{S}(x)w = \mathbf{R}(x)w$ for all $x \in G$ and $w \in W$, we say that S is a subrepresentation of R. Clearly, then there is a basis of C^n such that $\mathbf{R}(x)$ has the form

$$\mathbf{R}(x) = \begin{bmatrix} \mathbf{S}(x) & * \\ \mathbf{0} & * \end{bmatrix}.$$

A representation $\mathbf{R}(x)$ is called irreducible if its only subrepresentations are $\mathbf{R}$ and $\mathbf{0}$.

Consider the space of complex functions on G, i.e., $L = \{f | f : G \to C\}$ which is a vector space of dimension $n = |G|$ over C. Define the convolution product $*$

$$(f * g)(x) = \sum_{t \in G} f(t)g(xt^{-1}) = \sum_{t \in G} f(t^{-1}x)g(t).$$

It is straightforward to check that $*$ makes L into an algebra over C.

Definition 2.2 *Two representations* $\mathbf{S}$ *and* $\mathbf{R}$ *are called equivalent if there is a matrix* $\mathbf{T}$ *such that* $\mathbf{S} = \mathbf{T}^{-1}\mathbf{R}\mathbf{T}$.

2.1.1 Complete reducibility

Proposition 2.1 *Suppose* $\mathbf{S} : G \to U(m)$ *is a subrepresentation of* $\mathbf{R} : G \to U(n)$. *Then* $\mathbf{R}$ *is equivalent to the representation*

$$(\mathbf{S} \oplus \mathbf{V})(x) = \begin{bmatrix} \mathbf{S}(x) & \mathbf{0} \\ \mathbf{0} & \mathbf{V}(x) \end{bmatrix}.$$

Proof. Since $\mathbf{R}$ is unitary, it leaves the inner product $\langle,\rangle$ invariant. As $\mathbf{S}$ is a subrepresentation of $\mathbf{R}$ it follows that there is a subspace $U_1 \subseteq U$ such that

$$\mathbf{R}(x)U_1 \subseteq U_1,$$
$$\mathbf{S}(x)u = \mathbf{R}(x)u,$$

for all $x \in G, u \in U_1$.

Define the orthogonal complement $U_1^\perp$ of U_1 as

$$U_1^\perp = \{u \in U | \langle u, u_1 \rangle = 0 \text{ for all } u_1 \in U_1\}.$$

Then, $\mathbf{R}(x)U_1^\perp \subseteq U_1^\perp$ and $\mathbf{V}(x)$ defined as the restriction of $\mathbf{R}$ to $U_1^\perp$ is a subrepresentation of $\mathbf{R}$. Since $U = U_1 \oplus U_1^\perp$, the proposition follows.

By induction $\mathbf{R}(x)$ is equivalent to

$$(\mathbf{S}_1(x) \oplus \cdots \oplus \mathbf{S}_r(x)) = \begin{bmatrix} \mathbf{S}_1(x) & 0 & \cdots & 0 \\ 0 & \mathbf{S}_2(x) & \cdots & 0 \\ \cdots & \cdots & \cdots & \cdots \\ 0 & 0 & \cdots & \mathbf{S}_r(x) \end{bmatrix}$$

$$= \begin{bmatrix} \mathbf{S}_1(x) & & 0 \\ & \ddots & \\ 0 & & \mathbf{S}_r(x) \end{bmatrix}.$$

We present the famous result called Schur lemma that is very useful in establishing certain properties of representations.

Lemma 2.1 *(Schur)*
Let $\mathbf{R} : G \to U(n)$ *and* $\mathbf{S} : G \to U(m)$ *be two representations and define the space* $I(\mathbf{R}, \mathbf{S})$ *by*

$$I(\mathbf{R}, \mathbf{S}) = \{\Phi : C^m \to C^n | \Phi \text{ is linear and } \Phi \circ \mathbf{S}(x) = \mathbf{R}(x) \circ \Phi, \text{ for all } x \in G\},$$

that is, such linear operators that the following diagram is commutative for all x

$$
\begin{array}{ccc}
C^m & \xrightarrow{\ \Phi\ } & C^n \\
\downarrow{\scriptstyle S(x)} & & \downarrow{\scriptstyle R(x)} \\
C^m & \xrightarrow{\ \Phi\ } & C^n
\end{array}
$$

Then,

1. *The restriction of* $\mathbf{R}$ *to* $\mathrm{Ker}\Phi$ *is a subrepresentation of* $\mathbf{R}$, *and the restriction of* $\mathbf{S}$ *to* $\mathrm{Im}\Phi$ *is a subrepresentation of* $\mathbf{S}$.

2. *If* $\mathbf{R}$ *and* $\mathbf{S}$ *are irreducible and* $\Phi \neq 0$, *then* Φ *is an isomorphism.*

3. *If* $\Phi \in I(\mathbf{R}, \mathbf{R})$, *then there is* $\lambda \in C$ *such that* $\Phi v = \lambda v$ *for all* $v \in C^n$, *i.e.,* $\Phi = \lambda \mathbf{I}$.

4. *If* R *and* S *are both irreducible, then*

$$\dim I(\mathbf{R}, \mathbf{S}) = \begin{cases} 1, & \text{if } \mathbf{R}, \mathbf{S} \text{ are equivalent,} \\ 0 & \text{otherwise.} \end{cases}$$

Proof.

1. For $v \in \mathrm{Ker}\Phi$, we have $\Phi(\mathbf{R}(x)v) = \mathbf{S}(x)\Phi v = \mathbf{S}(x)0 = 0$. Thus $\mathbf{R}(x)v \in \mathrm{Ker}\Phi$ for all $x \in G$ and we have that $v \in \mathrm{Ker}\Phi$. Let $w \in \mathrm{Im}\Phi$, it follows

that $w = \Phi v$ for some $v \in C^m$. Therefore, $\mathbf{S}(x)w = \mathbf{S}(x)\Phi v = \Phi \mathbf{R}(x)v = \Phi(\mathbf{R}(x)v) \in \text{Im}\Phi$ implying that $\text{Im}\Phi$ is a subrepresentation of $\mathbf{S}$.

2. If $\Phi \neq 0$, it follows that $\text{Ker}\Phi \neq C^n$, and since $\mathbf{R}$ is irreducible, $\text{Ker}\Phi = \{0\}$. Now, $\Phi : C^m \to C^n$ is one to one and since S is irreducible, $\text{Im}\Phi = C^n$ and $m = n$.

3. Let λ be an eigenvalue of Φ and define $V_\lambda = \{v \in C^m | \Phi v = \lambda v\}$. Now, $V_\lambda \neq \{0\}$ and $\mathbf{R}(x)V_\lambda \subseteq V_\lambda$ for all x. Thus the restriction of $\mathbf{R}$ to V_λ is a subrepresentation. Since $\mathbf{R}$ irreducible, $V_\lambda = C^m$. Thus, $\Phi = \lambda \mathbf{I}$.

4. Consider $\Phi, \Psi \neq 0$, that belong to $I(\mathbf{R},\mathbf{S})$. By 2. they are isomorphisms and it follows that Φ satisfies $\Phi^{-1}\mathbf{S}(x) = \mathbf{R}(x)\Phi^{-1}$ for all $x \in G$.

$$
\begin{array}{ccc}
C^m & \xrightarrow{\Phi,\Psi} & C^n \\
\downarrow{\scriptstyle R(x)} & & \downarrow{\scriptstyle S(x)} \\
C^m & \xrightarrow{\Phi,\Psi} & C^n
\end{array}
$$

Thus, $(\Phi^{-1}\Psi)\mathbf{R}(x) = \Phi^{-1}\mathbf{S}(x)\Psi = \mathbf{R}(x)(\Phi^{-1}\Psi)$ for all $x \in G$ implying that $\Phi^{-1}\Psi \in I(\mathbf{R},\mathbf{R})$. By 3. we have $\Phi^{-1}\Psi = \lambda \mathbf{I}$, or equivalently $\Psi = \lambda \Phi$ and so $\dim I(\mathbf{R},\mathbf{S}) \leq 1$.

Notice that an irreducible representation of an Abelian group has degree one. This can be seen as follows.

Fix an element $x_0 \in G$ and consider the map $\Phi_{x_0} : C^n \to C^n$ where $\Phi_{x_0}v = \mathbf{R}(x_0)v$. Since G is an Abelian group, $\Phi_{x_0} \circ \mathbf{R}(x) = \mathbf{R}(x) \circ \Phi_{x_0}$ for all $x \in G$ implying that $\Psi_{x_0} \in I(\mathbf{R},\mathbf{R})$, whence $\mathbf{R}(x_0) = \Phi_{x_0} = \lambda \mathbf{I}$ for some λ. Thus, $\mathbf{R}(x_0)$ is diagonal and $\mathbf{R}$ can be irreducible only if $n = 1$.

Definition 2.3 *(Character of a representation)*
The character χ_R of a representation $\mathbf{R}$ is $\text{Tr}(\mathbf{R})$ *where* $\text{Tr}\mathbf{A}$ *means the trace of* $\mathbf{A}$.

Recall that the usual inner product in the space of square integrable functions on G into C, $L^2(G)$, is in the case of a finite group G

$$\langle f, g \rangle = \sum_{x \in G} f(x)\overline{g(x)}.$$

Theorem 2.1 *(Schur orthogonality relations)*

1. Let $\mathbf{R} : G \to U(n)$ and $\mathbf{S} : G \to U(m)$ be inequivalent and irreducible representations of a group G. Then,

$$\langle R_{ij}, S_{rs} \rangle = \sum_{x \in G} R_{ij}(x)\overline{S_{rs}(x)} = 0, \quad \text{for all} \quad i, j, r, s,$$

where $R_{ij}^{(n)}(x)$ denotes the i, j-th entry of the matrix $\mathbf{R}(x)$.

2. *In particular*

$$\langle R_{ij}, R_{rs} \rangle = r^{-1}|G|\delta_{ir}\delta_{js},$$

where r is the degree of $\mathbf{R}$.

3. *Let* $\mathbf{R}$ *and* $\mathbf{S}$ *be irreducible representations of* G. *Then,*

$$\langle \chi_R, \chi_S \rangle = \begin{cases} 0, & \text{if } \mathbf{R} \text{ and } \mathbf{S} \text{ are inequivalent,} \\ |G|, & \text{if } \mathbf{R} \text{ and } \mathbf{S} \text{ are equivalent.} \end{cases}$$

Proof.

1. Consider the matrix

$$\mathbf{P} = \sum_{x \in G} \mathbf{R}(x)\mathbf{C}\mathbf{S}^{-1}(x),$$

where $\mathbf{C}$ is an $(n \times m)$ matrix to be specified later. Now, $\mathbf{P} \in I(\mathbf{R}, \mathbf{S})$ since

$$\begin{aligned} \mathbf{R}(y)\mathbf{P} &= \sum_{x \in G} \mathbf{R}(yx)\mathbf{C}\mathbf{S}(x^{-1}) \\ &= \sum_{u \in G} \mathbf{R}(u)\mathbf{C}\mathbf{S}(u^{-1}y) = \mathbf{P}\mathbf{S}(y). \end{aligned}$$

Since $\mathbf{R}$ and $\mathbf{S}$ are irreducible and inequivalent, we have by claim 2. of Schur lemma that $\mathbf{P} = 0$. Choosing $\mathbf{C} = \mathbf{E}_{js}$, the matrix having 1 in the entry sj and 0 elsewhere, we see that

$$\mathbf{P}_{ir} = \sum_{x \in G} \mathbf{R}_{ij}(x)\mathbf{S}_{rs}(x) = 0.$$

2. Again, let $\mathbf{C}$ be any nonzero $n \times n$ matrix and write

$$\mathbf{P} = \sum_{x \in G} \mathbf{R}(x)\mathbf{C}\mathbf{R}^{-1}(x).$$

Now, by claim 3. of Schur lemma we have $\mathbf{P} = \lambda \mathbf{I}$. Setting $\mathbf{C} = \mathbf{E}_{ii}$ taking trace on both sides we obtain $\text{Tr}(\mathbf{P}) = \lambda n = \lambda r$. since trace is invariant under similarity transform we have

$$\text{Tr}\sum_{x \in G} \mathbf{R}(x)\mathbf{C}\mathbf{R}^{-1}(x) = \text{Tr}\sum_{x \in G} \mathbf{E}_{ii} = |G|.$$

Thus,

$$\lambda_C = r^{-1}|G|.$$

Setting now $\mathbf{C} = \mathbf{E}_{js}$, as above we obtain

$$P_{ir} = \sum_{x \in G} R_{ij}(x)\overline{R_{rs}(x)} = r^{-1}|G|\delta_{ir}\delta_{js}.$$

3. Let $\mathbf{R}$ and $\mathbf{S}$ be inequivalent and irreducible. By definition,

$$
\begin{aligned}
\langle \chi_R, \chi_S \rangle &= \sum_{x \in G} \sum_i R_{ii}(x) \sum_j \overline{S_{jj}(x)} \\
&= \sum_i \sum_j \sum_{x \in G} R_{ii}(x)\overline{S_{jj}(x)} = 0.
\end{aligned}
$$

If $\mathbf{R}$ and $\mathbf{S}$ are equivalent, by invariance of trace and claim 2,

$$\langle \chi_R, \chi_S \rangle = \langle \chi_R, \chi_R \rangle = \sum_i \sum_j \sum_{x \in G} R_{ii}(x)\overline{R_{jj}(x)} = rr^{-1}|G| = |G|.$$

Thus, we can say that if $\mathbf{R}$ and $\mathbf{S}$ are two representations of G and we have the direct sum representations

$$
\begin{aligned}
\mathbf{R} &\approx n_1\mathbf{R}_1 \oplus \cdots \oplus n_k\mathbf{R}_k, \\
\mathbf{S} &\approx m_1\mathbf{R}_1 \oplus \cdots \oplus m_k\mathbf{R}_k,
\end{aligned}
$$

where $\mathbf{R}_i$ are irreducible representations of G. Then,

$$\langle \chi_R, \chi_S \rangle = |G| \sum_{i=1}^k n_i m_i, \qquad (2.1)$$

and

$$\mathrm{dim}I(\mathbf{R}, \mathbf{S}) = |G|^{-1}\langle \chi_R, \chi_S \rangle.$$

Definition 2.4 *(Left regular representation)*
Let $x \in G$ and consider the permutation of elements of G, $y \to x^{-1}y$ and let $\mathbf{L}(x)$ be the corresponding $(|G| \times |G|)$ permutation matrix. The representation $\mathbf{L}(x)$ is called the left regular representation of G.

Lemma 2.2 *Let $\mathbf{L}(x)$ be the left regular representation of G. Then,*

$$\chi_L(x) = \begin{cases} |G|, & \text{if } x = e, \text{ the identity of } G, \\ 0, & \text{otherwise}, \end{cases}$$

and every irreducible representation $R \in \Gamma$ is contained in L with the multiplicity r. Thus, if $\Gamma = \{\mathbf{R}_1, \ldots, \mathbf{R}_k\}$, then L is similar to direct sum

$$L \approx (r_1)\mathbf{R}_1 \oplus \cdots \oplus (r_k)\mathbf{R}_k, \qquad (2.2)$$

and

$$\sum_{\mathbf{R}_i \in \Gamma} (r_i)^2 = |G|. \tag{2.3}$$

Proof. If $x \neq e$, then $\mathbf{L}(x)$ is a permutation matrix that fixes no element. Thus, $\mathrm{Tr}\mathbf{R}(x) = 0$. $\mathbf{R}(e)$ is the identity matrix and so $\mathrm{Tr}\mathbf{R}(e) = |G|$.

By (2.1), the multiplicity of $\mathbf{R}_i$ in L is found by

$$\frac{1}{|G|}\langle \chi_L, \chi_{\mathbf{R}_i} \rangle = \frac{1}{|G|} \sum_{y \in G} \chi_L(y)\overline{\chi_{\mathbf{R}_i}(y)},$$

$$= \frac{1}{|G|}|G|\chi_{\mathbf{R}_i}(e) = r_i.$$

(2.3) follows from (2.2).

2.2 FOURIER TRANSFORM ON FINITE GROUPS

Now we can collect the results above and define Fourier Analysis on Finite Groups.

Let G be a finite group and Γ the dual object, i.e., set of all inequivalent irreducible representations of G.

The matrix entries $R_{ij}(x)$ and as well $R_{ij}(x^{-1})$, of the representations $\mathbf{R} \in \Gamma$ as elements of $L^2(G)$ form a complete orthogonal set in $L^2(G)$. This statement is due to Weyl and his student Peter [13] and usually called Peter-Weyl theorem, see for example, [8]. As above, denote by r the degree of $\mathbf{R}$. By normalizing we find that $(r|G|^{-1})^{1/2}R_{ij}$, $\mathbf{R} \in \Gamma$, $1 \leq i, j \leq r$ form a complete orthonormal system in $L^2(G)$. Thus we can expand

$$f = \sum_{\mathbf{R} \in \Gamma} \sum_{1 \leq i,j \leq r} (r|G|^{-1})^{1/2}\langle f, R_{ij} \rangle (r|G|^{-1})^{1/2}R_{ij}$$

$$= |G|^{-1} \sum_{\mathbf{R} \in \Gamma} \sum_{1 \leq i,j \leq r} r\langle f, R_{ij} \rangle R_{ij}$$

and by Parseval relation

$$\|f\|^2 = \langle f, f \rangle = \sum_{\mathbf{R} \in \Gamma} \sum_{1 \leq i,j \leq r} r|\langle f, R_{ij} \rangle|^2.$$

These relations imply that it is reasonable to define a Fourier transform on functions $f \in L^2(G)$ in such a way that the structure of the non-Abelian group becomes expressed in the transform, e.g. such that the familiar relations of usual Fourier transform with regard to convolutions hold for convolutions defined by the non-Abelian group structure.

Definition 2.5 *(Fourier transform)*
Let $f : G \to C$ be a function, Γ the dual object of G, $\mathbf{R} \in \Gamma$ and denote by r the

degree of **R**. *The Fourier transform of f is defined by*

$$\mathcal{F}\{f\}(\mathbf{R}) = \mathbf{S}_f(\mathbf{R}) = r|G|^{-1} \sum_{x \in G} f(x)\mathbf{R}(x^{-1}).$$

Thus, the Fourier transform at representation **R** is an $(r \times r)$ matrix where r is the degree of **R**. By Peter-Weyl theorem the function f can be recovered by the inverse formula

$$
\begin{aligned}
f(x) &= \sum_{\mathbf{R} \in \Gamma} \sum_{1 \leq i,j \leq r} \mathbf{S}_f(\mathbf{R})R_{ij}(x) \\
&= \sum_{\mathbf{R} \in \Gamma} \mathrm{Tr}(\mathbf{S}_f(\mathbf{R})\mathbf{R}(\mathbf{x})).
\end{aligned}
$$

We will later give list of properties of Fourier transformations on finite groups. As an example, let us consider the convolution property. Let $f, g \in L^2(g)$. Then

$$r|G|^{-1}\mathcal{F}\{f * g\}(\mathbf{R}) = \mathcal{F}\{f\}(\mathbf{R})\mathcal{F}\{g\}(\mathbf{R}),$$

where the multiplication on the right-hand side is multiplication of $(r \times r)$ matrices. This can be seen as in the Abelian case.

$$
\begin{aligned}
r|G|^{-1}\mathcal{F}\{f * g\}(\mathbf{R}) &= r^2|G|^{-2} \sum_{x \in G}\left(\sum_{u \in G} f(u)g(xu^{-1})\right)\mathbf{R}(x^{-1}) \\
&= r^2|G|^{-2} \sum_{u \in G} f(u)\mathbf{R}(u^{-1}) \sum_{x \in G} g(xu^{-1})\mathbf{R}(ux^{-1}) \\
&= r|G|^{-1} \sum_{u \in G} f(u)\mathbf{R}(u^{-1}) \, r|G|^{-1} \sum_{v \in G} g(v)\mathbf{R}(v^{-1}).
\end{aligned}
$$

Here we will fix the notation that will be used later in the book. Let G be a finite, not necessarily Abelian, group of order $g = G$. We associate (permanently and bijectively) with each group element a non-negative integer from the set $\{0, 1, \ldots, g-1\}$, and 0 is associated with the group identity. Thus, each group element will be identified with the fixed non-negative integer associated with it and with no other element. We assume that G can be represented as a direct product of subgroups $G_1, \ldots, G_n$ of orders $g_1, \ldots, g_n$, respectively, i.e.,

$$G = \times_{i=1}^n G_i, \quad g = \Pi_{i=1}^n g_i, \quad g_1 \leq g_2 \leq \cdots \leq g_n. \tag{2.4}$$

The convention adopted above for the notation of group elements applies to the subgroups G_i as well. Provided that the notational bijections of the subgroups and of G are consistently chosen, each $x \in G$ can be uniquely represented as

$$x = \sum_{i=1}^n a_i x_i, \quad x_i \in G_i, \quad x \in G, \tag{2.5}$$

with

$$a_i = \begin{cases} \Pi^n_{j=i+1} g_j, & i = 1, \ldots, n-1 \\ 1, & i = n, \end{cases}$$

where g_j is the order of G_j and $0 \le x_i < g_i$, $i = 1, \ldots, n$.

The group operation $\circ$ of G can be expressed in terms of the group operations $\overset{\circ}{i}$ of the subgroups G_i, $i = 1, \ldots, n$ by

$$x \circ y = (x_1 \overset{\circ}{1} y_1, \ x_2 \overset{\circ}{2} y_2, \ \ldots \ x_n \overset{\circ}{n} y_n), \quad x, y \in G, \quad x_i, y_i \in G_i. \qquad (2.6)$$

Denote by P the complex field or a finite field and by $P(G)$ the space of functions f mapping G into P, i.e., $f : G \to P$. Due to the assumption (2.4) and the relation (2.5), each function $f \in P(G)$ can be considered as an n-variable function $f(x_1, \ldots, x_n)$, $x_i \in G_i$.

We denote by K the cardinality of the dual object Γ or $\Gamma(G)$ i.e. number of equivalence classes of irreducible representations of G over P. Each such equivalence class contains just one unitary representation. We shall denote the elements of Γ in some fixed order by $\mathbf{R}_0, \mathbf{R}_1, \ldots, \mathbf{R}_{K-1}$ and by $\mathbf{R}_w(x)$ the value of $\mathbf{R}$ at $x \in G$. Because we now use subscripts to identify the representations, we, from now on, will use superscripts for the elements of the matrices of the representations e.g. $R_w^{(i,j)}(x)$, $i, j = 1, 2, \ldots, r_w$.

Notice that group representations may be used in different orders when this is more appropriate for particular applications. This is the same situation as in the case of Abelian groups , for example, the dyadic group where the group characters are Walsh functions [20], which are used in different orderings [9].

If the group G is representable in the form (2.4), then its unitary irreducible representations can be obtained as Kronecker products of the unitary irreducible representations of subgroups G_i, $i = 1, \ldots, n$. Therefore, the number K of unitary irreducible representations of G is

$$K = \prod_{i=1}^{n} K_i, \qquad (2.7)$$

where K_i is the number of unitary irreducible representations of the subgroup G_i.

Now, for a given group G of the form (2.4), the index w of each unitary irreducible representation $\mathbf{R}_w$ can be written as:

$$w = \sum_{i=1}^{n} b_i w_i, \quad w_i \in \{0, 1, \ldots, K_i - 1\}, \quad w \in \{0, 1, \ldots, K - 1\},$$

with

$$b_i = \begin{cases} \prod_{j=i+1}^{n} K_j, & i = 1, \ldots, n-1, \\ 1, & i = n, \end{cases} \qquad (2.8)$$

where K_j is the number of unitary irreducible representations of the subgroup G_j.

Table 2.1   Group operation of S_3.

∘	0	1	2	3	4	5
0	0	1	2	3	4	5
1	1	2	0	5	3	4
2	2	0	1	4	5	3
3	3	4	5	0	1	2
4	4	5	3	2	0	1
5	5	3	4	1	2	0

We denote by $\mathbf{W}_i$ the vector of length K of the values of w_i considered as an n-variable discrete function.

As noted above, the functions $R_w^{(i,j)}(x)$, $w = 0, 1, \ldots, K-1$, $i, j = 1, \ldots, r_w$ form an orthogonal system in the space $P(G)$ and the direct and the inverse Fourier transform of a function $f \in P(G)$ are defined respectively by

$$\mathbf{S}_f(w) = r_w g^{-1} \sum_{u=0}^{g-1} f(u) \mathbf{R}_w(u^{-1}), \qquad (2.9)$$

$$f(x) = \sum_{w=0}^{K-1} Tr(\mathbf{S}_f(w) \mathbf{R}_w(x)). \qquad (2.10)$$

Here and in the sequel we shall assume, without explicitly saying so, that all arithmetical operations are carried out in the field P.

Note that if in (2.4), there are two identical non-Abelian subgroups $G_i = G_j, i \neq j$, then the Kronecker product of unitary irreducible representations of the subgroups does not produce the unitary irreducible representations of G. However, the transform defined in terms of so generated a set of linearly independent functions is still denoted as the (generalized) Fourier transform, with the generalization achieved through the formal application of the decomposition used in FFT-like algorithms, see, for example, [14].

Example 2.1 *Let $S_3 = (0, (132), (123), (12), (13), (23), \circ)$ be the symmetric group of permutations of order 3. According to the convention adopted in this book, the group elements of S_3 will be denoted by 0,1,2,3,4,5, respectively. Using this notation the group operation of S_3 is shown in Table 2.1. The unitary irreducible representations of S_3 over C are given in Table 2.2. The group characters and the set $R_w^{(i,j)}$ of functions representing a basis in the space of complex functions on S_3 are given in Table 2.3 and Table 2.4.*

Table 2.2 The unitary irreducible representations of S_3 over C.

x	R_0	R_1	R_2
0	1	1	**I**
1	1	1	**A**
2	1	1	**B**
3	1	-1	$\cdot$ **C**
4	1	-1	**D**
5	1	-1	**E**

$$\mathbf{I} = \begin{bmatrix} 1 & 0 \\ 0 & 1 \end{bmatrix}, \qquad \mathbf{A} = -2^{-1}\begin{bmatrix} 1 & -\sqrt{3} \\ \sqrt{3} & 1 \end{bmatrix}, \qquad \mathbf{B} = -2^{-1}\begin{bmatrix} 1 & \sqrt{3} \\ -\sqrt{3} & 1 \end{bmatrix},$$

$$\mathbf{C} = \begin{bmatrix} 1 & 0 \\ 0 & -1 \end{bmatrix}, \qquad \mathbf{D} = -2^{-1}\begin{bmatrix} 1 & -\sqrt{3} \\ -\sqrt{3} & -1 \end{bmatrix}, \qquad \mathbf{E} = -2^{-1}\begin{bmatrix} 1 & \sqrt{3} \\ \sqrt{3} & -1 \end{bmatrix},$$

Table 2.3 The group characters of S_3 over C.

x	χ_0	χ_1	χ_2
0	1	1	2
1	1	1	2
2	1	1	2
3	1	-1	0
4	1	-1	0
5	1	-1	0

Table 2.4 The set $R_w^{(i,j)}(x)$ of S_3 over C.

x	R_0	R_1	$R_2^{(0,0)}$	$R_2^{(0,1)}$	$R_2^{(1,0)}$	$R_2^{(1,1)}$
0	1	1	1	0	0	1
1	1	1	$-\frac{1}{2}$	$\frac{\sqrt{3}}{2}$	$-\frac{\sqrt{3}}{2}$	$-\frac{1}{2}$
2	1	1	$-\frac{1}{2}$	$-\frac{\sqrt{3}}{2}$	$\frac{\sqrt{3}}{2}$	$-\frac{1}{2}$
3	1	-1	1	0	0	-1
4	1	-1	$-\frac{1}{2}$	$\frac{\sqrt{3}}{2}$	$\frac{\sqrt{3}}{2}$	$\frac{1}{2}$
5	1	-1	$-\frac{1}{2}$	$-\frac{\sqrt{3}}{2}$	$-\frac{\sqrt{3}}{2}$	$\frac{1}{2}$

Table 2.5 Unitary irreducible representations of S_3 over $GF(11)$.

x	$\mathbf{R_0}$	$\mathbf{R_1}$	$\mathbf{R_2}$
0	1	1	$\mathbf{I}$
1	1	1	$\mathbf{A}$
2	1	1	$\mathbf{B}$
3	1	10	$\mathbf{C}$
4	1	10	$\mathbf{D}$
5	1	10	$\mathbf{E}$

$$\mathbf{I} = \begin{bmatrix} 1 & 0 \\ 0 & 1 \end{bmatrix} \quad \mathbf{A} = \begin{bmatrix} 5 & 8 \\ 3 & 5 \end{bmatrix} \quad \mathbf{B} = \begin{bmatrix} 5 & 3 \\ 8 & 5 \end{bmatrix}$$

$$\mathbf{C} = \begin{bmatrix} 1 & 0 \\ 0 & 10 \end{bmatrix} \quad \mathbf{D} = \begin{bmatrix} 5 & 8 \\ 8 & 6 \end{bmatrix} \quad \mathbf{E} = \begin{bmatrix} 5 & 3 \\ 3 & 6 \end{bmatrix}$$

Example 2.2 *The unitary irreducible representations of S_3 over the Galois field $GF(11)$ are shown in Table 2.5. The group characters and the basis $R_w^{(i,j)}$ of S_3 over $GF(11)$ are given in Table 2.6 and Table 2.7.*

2.3 PROPERTIES OF THE FOURIER TRANSFORM

The main properties of the Fourier transform on finite non-Abelian groups are analogous to those of the Fourier transform on Abelian groups, such as for example, the

Table 2.6 The group characters of S_3 over $GF(11)$.

x	χ_0	χ_1	χ_2
0	1	1	2
1	1	1	10
2	1	1	10
3	1	10	0
4	1	10	0
5	1	10	0

Table 2.7 The set $R_w^{(i,j)}(x)$ of S_3 over $GF(11)$.

x	R_0	R_1	$R_2^{(0,0)}$	$R_2^{(0,1)}$	$R_2^{(1,0)}$	$R_2^{(1,1)}$
0	1	1	1	0	0	1
1	1	1	5	8	3	5
2	1	1	5	3	8	5
3	1	10	1	0	0	10
4	1	10	5	8	8	6
5	1	10	5	3	3	6

Walsh transform [1], the Vilenkin-Chrestenson transform [4], [19], see also [9], [12], and in particular to that of the Fourier transform on the real line [3].

Theorem 2.2 *The main properties of the Fourier transform on finite non-Abelian groups are the following:*

1. *Linearity: For all* $\alpha_1, \alpha_2 \in P$, $f_1, f_2 \in C(G)$,

$$\mathbf{S}_{\alpha_1 f_1 + \alpha_2 f_2}(w) = \alpha_1 \mathbf{S}_{f_1}(w) + \alpha_2 \mathbf{S}_{f_2}(w).$$

2. *Right group translation: For all* $\tau \in G$,

$$\mathbf{S}_{f(x\tau)}(w) = \mathbf{R}_w(\tau)\mathbf{S}_f(w).$$

3. *Group convolution: For two functions* $f_1, f_2 \in C(G)$ *the convolution is defined by*

$$(f_1 * f_2)(\tau) = \sum_{x \in G} f_1(x) f_2(\tau^{-1}x).$$

4. *Relative to the so defined convolution, the Fourier transform exhibits the following property*

$$r_w g^{-1} \mathbf{S}_{(f_1 * f_2)(\tau)}(w) = \mathbf{S}_{f_1}(w)\mathbf{S}_{f_2}(w).$$

It should be noted that unlike the Fourier transform on Abelian groups, a dual statement cannot be formulated since the dual object Γ does not exhibit a group structure suitable for definition of a convolution of functions on Γ.

5. *Parseval theorem: For all* $f_1, f_2 \in P(G)$,

$$\sum_{x \in G} f_1(x)\overline{f}_2(x) = g \sum_{\mathbf{R}_w \in \Gamma(G)} r_w^{-1} \mathrm{Tr}(\mathbf{S}_{f_1}(w)\mathbf{S}_{f_2}^*(w)),$$

where $\overline{f}$ denotes the complex-conjugate of f, $\mathbf{S}_{f_2}^*(\cdot)$ is the conjugate transpose of $\mathbf{S}_{f_2}(\cdot)$, i.e., $\mathbf{S}_{f_2}^*(\cdot) = (\overline{\mathbf{S}}_{f_2}(\cdot))^T$.

6. *The Wiener-Khinchin theorem: For two functions* $f_1, f_2 \in P(G)$, *the cross-correlation function is defined by*

$$R_{f_1, f_2(\tau)} = \sum_{x \in G} f_1(x)\overline{f_2(x\tau^{-1})}.$$

The autocorrelation function is the cross-correlation function for $f_1 = f_2$.

Denote by $\mathbf{F}_G$ and $\mathbf{F}_G^{-1}$ the direct and inverse Fourier transform on G, respectively, and by $\mathbf{F}_G^*$ the transform such that

$$(\mathbf{F}_G^*(f))(w) = \mathbf{S}_f^*(w).$$

With this notation the Wiener-Khinchin theorem on G is defined by

$$B_{f_1,f_2} = gF_G^{-1}(r_w^{-1}\mathbf{F}_G(f_1)\mathbf{F}_G^*(f_2)).$$

Proof. Properties 1, 2 and 3 follow immediately from the definition of the Fourier transform and its inverse.

The Parseval theorem can be proved by using the orthogonality and the unitarity of $\mathbf{R}_w(\cdot)$, i.e.,

$$g \sum_{\mathbf{R}_w \in \Gamma(G)} r_w^{-1} \mathrm{Tr}(\mathbf{S}_{f_1}(w)\mathbf{S}_{f_2}^*(w))$$

$$= g^{-1} \sum_{\mathbf{R}_w \in \Gamma((G)} r_w \mathrm{Tr}((\sum_{x_1 \in G} f_1(x_1)\mathbf{R}_w(x_1^{-1}))(\sum_{x_2 \in G} \overline{f_2(x_2)}\mathbf{R}_w^*(x_2))$$

$$= g^{-1} \sum_{x_1,x_2 \in G} f_1(x_1)\overline{f_2(x_2)} \sum_{\mathbf{R}_w \in \Gamma((G)} r_w \mathrm{Tr}(x_1^{-1}x_2)$$

$$= \sum_{x \in G} f_1(x)\overline{f_2(x)}.$$

The Wiener-Khinchin theorem follows from the unitarity of $\mathbf{R}_w(x)$, the definition of the Fourier transform and the convolution theorem.

2.4   MATRIX INTERPRETATION OF THE FOURIER TRANSFORM ON FINITE NON-ABELIAN GROUPS

In our further consideration we need the generalized matrix multiplications defined as follows.

Definition 2.6 *Let* $\mathbf{A}$ *be an* $(m \times n)$ *matrix with elements* $a_{ij} \in P$, $i \in \{0, 1, \ldots, m-1\}$, $j \in \{0, 1, \ldots, n-1\}$. *Let* $[\mathbf{B}]$ *be an* $(n \times r)$ *matrix whose elements* $\mathbf{b}_{jk}$, $j \in \{0, \ldots, n-1\}$, $k \in \{0, 1, \ldots, r-1\}$ *are* $(p \times q)$ *matrices with elements in* P. *These matrices have the same order within a column, and may have different order in different columns of* $[\mathbf{B}]$. *We define the product* $\mathbf{A} \odot [\mathbf{B}]$ *as the* $(m \times r)$ *matrix* $[\mathbf{Y}]$ *whose elements* $\mathbf{y}_{ik}$, $i \in \{0, 1, \ldots, m-1\}$, $k \in \{0, 1, \ldots, r-1\}$ *are* $(p \times q)$ *matrices with elements in* P *given by*

$$\mathbf{y}_{ik} = \sum_{j=0}^{n-1} a_{ij}\mathbf{b}_{jk}.$$

The product $[\mathbf{B}] \odot \mathbf{A}$ is defined similarly.

Definition 2.7 *Let* $[\mathbf{Z}]$ *be an* $(m \times n)$ *matrix whose elements* $\mathbf{z}_{ij}$ $i \in \{0, 1, \ldots, m-1\}$, $j \in \{0, 1, \ldots, n-1\}$ *are square matrices of not necessarily mutually equal orders with elements in* P. *Let* $[\mathbf{B}]$ *be* $(n \times r)$ *matrix whose elements* $\mathbf{b}_{jk}$, $j \in$

$\{0, 1, \ldots, n-1\}$, $k \in \{0, 1, \ldots, r-1\}$ *are square matrices with elements in* P. *These matrices have the same order within a column, and may have different orders in different columns of* $[\mathbf{Z}]$ *and* $[\mathbf{B}]$. *Under the condition that the matrices* $\mathbf{z}_{ij}$ *and* $\mathbf{b}_{jk}$ *are of the same order or, if not, that one of them is of the order 1, the product of matrices* $[\mathbf{Z}]$ *and* $[\mathbf{B}]$ *is defined as the* $(m \times r)$ *matrix* $\mathbf{Y} = [\mathbf{Z}] \circ [\mathbf{B}]$ *whose elements* $y_{ik} \in P$ *are given by*

$$y_{ik} = \sum_{j=0}^{n-1} Tr(\mathbf{z}_{ij} \mathbf{b}_{jk}).$$

Definition 2.8 *Let* $[\mathbf{Z}]$ *be an* $(m \times n)$ *matrix whose elements* $\mathbf{z}_{ij}$, $i \in \{0, 1, \ldots, m-1\}$, $j \in \{0, 1, \ldots, n-1\}$ *are* $(p \times q)$ *matrices with elements in* P. *Let* $[\mathbf{B}]$ *be an* $(n \times r)$ *matrix whose elements* $\mathbf{b}_{jk}$, $j \in \{0, 1, \ldots, n-1\}$, $k \in \{0, 1, \ldots, r-1\}$ *are* $(s \times t)$ *matrices with elements in* P. *Both* $\mathbf{z}_{ij}$ *and* $\mathbf{b}_{jk}$ *are matrices that have the same order within a column, and may have different orders in different columns. The elementwise Kronecker product of matrices* $[\mathbf{Z}]$ *and* $[\mathbf{B}]$ *is defined as the* $(m \times r)$ *matrix* $[\mathbf{V}] = [\mathbf{Z}] \overset{\circ}{\otimes} [\mathbf{B}]$ *whose elements* $\mathbf{v}_{ik}$ *are given by*

$$\mathbf{v}_{ik} = \sum_{j=0}^{n-1} \mathbf{z}_{ij} \otimes \mathbf{b}_{jk},$$

where $\otimes$ *denotes the ordinary Kronecker product.*

The generalized matrix multiplication concepts, needed to describe the Fourier transform and its inverse on finite non-Abelian groups, were introduced in [15], [16]. Although we have never met these definitions, we cannot be sure that they are not to be found in the voluminous literature on matrix calculus.

By using the matrix operations thus introduced, the Fourier transform pair defined by (2.9) and (2.10) can be expressed as follows.

Let $f \in P(G)$ be given as a vector $\mathbf{f} = [f(0), \ldots, f(g-1)]^T$. Then its Fourier transform is given by

$$[\mathbf{S}_f] = g^{-1}[\mathbf{R}]^{-1} \odot \mathbf{f}, \tag{2.11}$$

where $[\mathbf{S}_f] = [\mathbf{S}_f(0), \ldots, \mathbf{S}_f(K-1)]^T$, and $[\mathbf{R}]^{-1} = [\mathbf{b}_{sq}]$ with $\mathbf{b}_{sq} = r_w \mathbf{R}_s^{-1}(q)$, $s \in \{0, 1, \ldots, K-1\}$, $q \in \{0, 1, \ldots, g-1\}$.

The inverse Fourier transform is given by

$$\mathbf{f} = [\mathbf{R}] \circ [\mathbf{S}_f], \tag{2.12}$$

where $[\mathbf{R}] = [\mathbf{a}_{ij}]$ with $\mathbf{a}_{ij} = \mathbf{R}_j(i)$, $i \in \{0, 1, \ldots, g-1\}, j \in \{0, 1, \ldots, K-1\}$.

2.5   FAST FOURIER TRANSFORM ON FINITE NON-ABELIAN GROUPS

The formulation of the fast Fourier transform on a finite decomposable group G of the form (2.4) is based on the consideration of the Fourier transform on G as n-dimensional Fourier transform each of them relative to one of the n subgroups G_j of G.

In this setting the Fourier transform can be written as:

$$
\begin{aligned}
\mathbf{S}_f(w) &= \mathbf{S}_f(w_1, \ldots, w_n) \\
&= r_w g^{-1} \sum_{x_1} \sum_{x_2} \cdots \sum_{x_n} f(x_1, x_2, \ldots, x_n) \bigotimes_{j=1}^{n} \mathbf{R}_{w_j}(x_j^{-1}) \\
&= r_w g^{-1} \sum_{x_n} (\ldots (\sum_{x_2} (\sum_{x_1} (f(x_1, x_2, \ldots, x_n) \mathbf{R}_{w_1}(x_1^{-1})) \\
&\quad \otimes \mathbf{R}_{w_2}(x_2^{-1})) \otimes \ldots) \otimes \mathbf{R}_{w_n}(x_n^{-1})).
\end{aligned}
$$

It follows that the Fourier transform can be performed in n steps defined as follows [10].

Step 1

$$
f_1(x_1, \ldots, x_n) = f(x_1, \ldots, x_n),
$$
$$
\mathbf{f}_2(w_1, x_2, \ldots, x_n) = \sum_{x_1} f_1(x_1, \ldots x_n) \mathbf{R}_{w_1}(x_1^{-1}).
$$

Step 2

$$
\mathbf{f}_3(w_1, w_2, x_3, \ldots, x_n) = \sum_{x_2} \mathbf{f}_2(w_1, x_2, \ldots x_n) \otimes \mathbf{R}_{w_2}(x_2^{-1}).
$$

Step j

$$
\mathbf{f}_j(w_1, \ldots, w_j, x_{j+1}, \ldots, x_n) = \sum_{x_j} \mathbf{f}_{j-1}(w_1, \ldots, w_{j-1}, x_j, \ldots x_n) \otimes \mathbf{R}_{w_j}(x_j^{-1}).
$$

Step n

$$
\mathbf{f}_{n+1}(w_1, \ldots, w_n) = \sum_{x_n} \mathbf{f}_n(w_1, \ldots x_n) \otimes \mathbf{R}_{w_n}(x_n^{-1}),
$$
$$
\mathbf{S}_f(w_1, \ldots, w_n) = r_w g^{-1} \mathbf{f}_{n+1}(w_1, \ldots, w_n).
$$

During the step j, the variables $w_1, \ldots, w_{j-1}, x_{j+1}, \ldots, x_n$ are fixed and the summation is performed through the variable $x_j \in G_j$.

The algorithm is probably best explained by an example.

Example 2.3 *Let G be the Quaternion (non-Abelian) group Q_2 of order 8. This group has two generators a and b and the group identity is denoted by e. If the group*

Table 2.8 Group operation for the quaternion group Q_2.

∘	0	1	2	3	4	5	6	7
0	0	1	2	3	4	5	6	7
1	1	2	3	0	5	6	7	4
2	2	3	0	1	6	7	4	5
3	3	0	1	2	7	4	5	6
4	4	7	6	5	2	3	0	1
5	5	4	7	6	3	2	1	0
6	6	5	4	7	0	3	2	1
7	7	6	5	4	1	0	3	2

operation is written as abstract multiplication, the following relations hold for the group generators: $b^2 = a^2, bab^{-1} = a^{-1}, a^4 = e$. If the following bijection V is chosen

x	e	a	a^2	a^3	b	ab	a^2b	a^3b
$V(x)$	0	1	2	3	4	5	6	7

then the full group operation is described in Table 2.8.

All the irreducible unitary representations over the complex field C are given in Table 2.9.

For notation convenience, the entries of column $\mathbf{R}_4$ can be denoted as

$$\mathbf{I}, \quad i\mathbf{A}, \quad -\mathbf{I}, \quad i\mathbf{B}, \quad \mathbf{C} \quad -i\mathbf{D} \quad \mathbf{E} \quad i\mathbf{D},$$

where

$$\mathbf{I} = \begin{bmatrix} 1 & 0 \\ 0 & 1 \end{bmatrix} \quad \mathbf{A} = \begin{bmatrix} 1 & 0 \\ 0 & -1 \end{bmatrix} \quad \mathbf{B} = \begin{bmatrix} -1 & 0 \\ 0 & 1 \end{bmatrix}$$

$$\mathbf{C} = \begin{bmatrix} 0 & -1 \\ 1 & 0 \end{bmatrix} \quad \mathbf{D} = \begin{bmatrix} 0 & 1 \\ 1 & 0 \end{bmatrix} \quad \mathbf{E} = \begin{bmatrix} 0 & 1 \\ -1 & 0 \end{bmatrix}$$

The dual object Γ of Q_2 is of order 5, since there are five irreducible unitary representations of this group.

Four of the representations are 1-dimensional and one is 2-dimensional. The Fourier transform matrix is

$$[\mathbf{Q}]^{-1} = \frac{1}{8} \begin{bmatrix} 1 & 1 & 1 & 1 & 1 & 1 & 1 & 1 \\ 1 & -1 & 1 & -1 & 1 & -1 & 1 & -1 \\ 1 & 1 & 1 & 1 & -1 & -1 & -1 & -1 \\ 1 & -1 & 1 & -1 & -1 & 1 & -1 & 1 \\ 2\mathbf{I} & 2i\mathbf{B} & -2\mathbf{I} & 2i\mathbf{A} & 2\mathbf{E} & 2i\mathbf{D} & 2\mathbf{C} & -2i\mathbf{D} \end{bmatrix},$$

Table 2.9   Irreducible unitary representations of Q_2 over C.

x	$\mathbf{R}_0$	$\mathbf{R}_1$	$\mathbf{R}_2$	$\mathbf{R}_3$	$\mathbf{R}_4$
0	1	1	1	1	$\begin{bmatrix} 1 & 0 \\ 0 & 1 \end{bmatrix}$
1	1	-1	1	-1	$\begin{bmatrix} i & 0 \\ 0 & -i \end{bmatrix}$
2	1	1	1	1	$\begin{bmatrix} -1 & 0 \\ 0 & -1 \end{bmatrix}$
3	1	-1	1	-1	$\begin{bmatrix} -i & 0 \\ 0 & i \end{bmatrix}$
4	1	1	-1	-1	$\begin{bmatrix} 0 & -1 \\ 1 & 0 \end{bmatrix}$
5	1	-1	-1	1	$\begin{bmatrix} 0 & -i \\ -i & 0 \end{bmatrix}$
6	1	1	-1	-1	$\begin{bmatrix} 0 & 1 \\ -1 & 0 \end{bmatrix}$
7	1	-1	-1	1	$\begin{bmatrix} 0 & i \\ i & 0 \end{bmatrix}$
	$r_0 = 1$	$r_1 = 1$	$r_2 = 1$	$r_3 = 1$	$r_4 = 2$

where the notation is as in Table 2.9.

Therefore, the Fourier spectrum of a function f on Q_2 consists of five coefficients, four 1-dimensional and one 2-dimensional and can be represented as a vector

$$[\mathbf{S}_f] = \left[\ \mathbf{S}_f(0)\quad \mathbf{S}_f(1)\quad \mathbf{S}_f(2)\quad \mathbf{S}_f(3)\quad \mathbf{S}_f(4)\ \right]^T.$$

For example, the Fourier spectrum of the function f on Q_2 given by the truth-vector $\mathbf{f} = [0\alpha00\beta\lambda00]^T$ *is given by*

$$[\mathbf{S}_f] = [\mathbf{Q}]^{-1} \circ \mathbf{f}$$

$$= \frac{1}{8}\begin{bmatrix} 1 & 1 & 1 & 1 & 1 & 1 & 1 & 1 \\ 1 & -1 & 1 & -1 & 1 & -1 & 1 & -1 \\ 1 & 1 & 1 & 1 & -1 & -1 & -1 & -1 \\ 1 & -1 & 1 & -1 & -1 & 1 & -1 & 1 \\ 2\mathbf{I} & 2i\mathbf{B} & -2\mathbf{I} & 2i\mathbf{A} & 2\mathbf{E} & 2i\mathbf{D} & 2\mathbf{C} & -2i\mathbf{D} \end{bmatrix} \circ \begin{bmatrix} 0 \\ \alpha \\ 0 \\ 0 \\ \beta \\ \lambda \\ 0 \\ 0 \end{bmatrix}$$

$$= \frac{1}{8}\begin{bmatrix} \alpha+\beta+\lambda \\ -\alpha+\beta-\lambda \\ \alpha-\beta-\lambda \\ -\alpha-\beta+\lambda \\ 2\begin{bmatrix} -i\alpha & -\beta+i\lambda \\ \beta+i\lambda & i\alpha \end{bmatrix} \end{bmatrix}$$

This function f is reconstructed from the spectrum $[\mathbf{S}_f]$ as

$$\mathbf{f} = [\mathbf{Q}] \circ [\mathbf{S}_f]$$

$$= \begin{bmatrix} 1 & 1 & 1 & 1 & \mathbf{I} \\ 1 & -1 & 1 & -1 & i\mathbf{A} \\ 1 & 1 & 1 & 1 & -\mathbf{I} \\ 1 & -1 & 1 & -1 & i\mathbf{B} \\ 1 & 1 & -1 & -1 & \mathbf{C} \\ 1 & -1 & -1 & 1 & -i\mathbf{D} \\ 1 & 1 & -1 & -1 & \mathbf{E} \\ 1 & -1 & -1 & 1 & i\mathbf{D} \end{bmatrix} \circ \frac{1}{8}\begin{bmatrix} \alpha+\beta+\lambda \\ -\alpha+\beta-\lambda \\ \alpha-\beta-\lambda \\ -\alpha-\beta+\lambda \\ 2\begin{bmatrix} -i\alpha & -\beta+i\lambda \\ \beta+i\lambda & i\alpha \end{bmatrix} \end{bmatrix}$$

$$= \frac{1}{8}\begin{bmatrix} (\alpha+\beta+\lambda)+(-\alpha+\beta-\lambda)+(\alpha-\beta-\lambda)+(-\alpha-\beta+\lambda)+\sigma_0 \\ (\alpha+\beta+\lambda)-(-\alpha+\beta-\lambda)+(\alpha-\beta-\lambda)-(-\alpha-\beta+\lambda)+\sigma_1 \\ (\alpha+\beta+\lambda)+(-\alpha+\beta-\lambda)+(\alpha-\beta-\lambda)+(-\alpha-\beta+\lambda)+\sigma_2 \\ (\alpha+\beta+\lambda)-(-\alpha+\beta-\lambda)+(\alpha-\beta-\lambda)-(-\alpha-\beta+\lambda)+\sigma_3 \\ (\alpha+\beta+\lambda)+(-\alpha+\beta-\lambda)-(\alpha-\beta-\lambda)-(-\alpha-\beta+\lambda)+\sigma_4 \\ (\alpha+\beta+\lambda)-(-\alpha+\beta-\lambda)-(\alpha-\beta-\lambda)+(-\alpha-\beta+\lambda)+\sigma_5 \\ (\alpha+\beta+\lambda)+(-\alpha+\beta-\lambda)-(\alpha-\beta-\lambda)-(-\alpha-\beta+\lambda)+\sigma_6 \\ (\alpha+\beta+\lambda)-(-\alpha+\beta-\lambda)-(\alpha-\beta-\lambda)+(-\alpha-\beta+\lambda)+\sigma_7 \end{bmatrix}$$

$$= [0,\alpha,0,0,\beta,\lambda,0,0]^T,$$

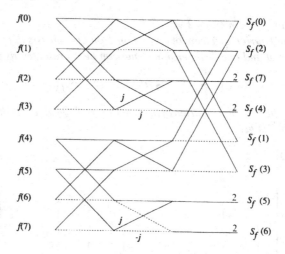

Fig. 2.1 FFT on the quaternion group Q_2.

where

$$
\begin{aligned}
\sigma_0 &= Tr(\mathbf{IS}_f(4)) = 0 \\
\sigma_1 &= Tr(i\mathbf{AS}_f(4)) = 4\alpha \\
\sigma_2 &= Tr(-\mathbf{IS}_f(4)) = 0 \\
\sigma_3 &= Tr(i\mathbf{BS}_f(4)) = -4\alpha \\
\sigma_4 &= Tr(\mathbf{CS}_f(4)) = \beta \\
\sigma_5 &= Tr(-i\mathbf{DS}_f(4)) = 4\lambda \\
\sigma_6 &= Tr(\mathbf{ES}_f(4)) = -4\beta \\
\sigma_7 &= Tr(i\mathbf{DS}_f(4)) = -4\lambda.
\end{aligned}
$$

Direct computation of the Fourier transform requires 64 operations for the quaternion group $G = Q_2$. Using the fast Fourier transform it can be computed by using 20 additions. The multiplications by the complex unity i are not considered. The corresponding flow-graph is shown in Figure 2.1.

Figure 2.2 shows the flow-graph of the FFT algorithm to calculate the inverse Fourier transform on Q_2.

The quaternion group Q_2 is a group structure which can be considered as the domain of signals defined on a set X_8 of eight elements. Using the mapping

$$
x = \sum_{i=0}^{2} x_i 2^{3-i}, \quad x_i \in \{0, 1\},
$$

the same set can be equipped with the structure of the dyadic group of order 8.

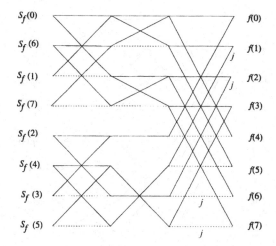

Fig. 2.2 Flow-graph for FFT algorithm for the inverse Fourier transform on Q_8.

Table 2.10 The discrete Walsh functions $\mathrm{wal}(i, x)$.

i, x	0	1	2	3	4	5	6	7
0	1	1	1	1	1	1	1	1
1	1	-1	1	-1	1	-1	1	-1
2	1	1	-1	-1	1	1	-1	-1
3	1	-1	-1	1	1	-1	-1	1
4	1	1	1	1	-1	-1	-1	-1
5	1	-1	1	-1	-1	1	-1	1
6	1	1	-1	-1	-1	-1	1	1
7	1	-1	-1	1	-1	1	1	-1

Recall that the dyadic group of order 2^n consists of the set of binary n-tuplex $x = (x_1, \ldots, x_n)$, $x_i \in \{0, 1\}$, under the componentwise addition modulo 2. Discrete Walsh functions, the discrete version of Walsh functions are the characters of the dyadic groups [7] and, therefore, form a basis in the space of complex functions on X_8. For $n = 3$ they are given in Table 2.10.

The algorithm for the computation of the Fourier transform on Q_2 can be compared to the algorithm for computation of the Fourier transform on the dyadic group of order 8 shown in Figure 2.3. The number of additions and subtractions to compute the Fourier transform on this group is 24 compared to 20 operations on Q_2 and 4 multiplications by the complex unity. The same algorithm can be used to calculate

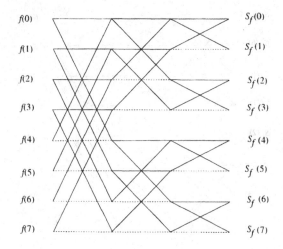

$f(0)$ $S_f(0)$
$f(1)$ $S_f(1)$
$f(2)$ $S_f(2)$
$f(3)$ $S_f(3)$
$f(4)$ $S_f(4)$
$f(5)$ $S_f(5)$
$f(6)$ $S_f(6)$
$f(7)$ $S_f(7)$

Fig. 2.3   FFT on the dyadic group of order 8.

the inverse Walsh transform, since it is a self-inverse transform up to the multiplicative constant 2^{-n}.

A method for an optimal implementation of the Fourier transform on finite not necessarily Abelian groups in a multiprocessor environment is presented in [14].

It is clear that the ordering of the subgroups G_j, indicated in (2.4), is not essential from the theoretical point of view. However, this order seems logical in some sense, and in our experience not only notationally convenient, but also useful in some practical considerations. Moreover, as is documented in [14], for a fixed set of constituent subgroups G_j their ordering as is required in (2.4) minimizes the number of data transfers in the implementation of the fast Fourier transform on G in a multiprocessor environment.

REFERENCES

1. Beauchamp, K.G., *Applications of Walsh and Related Functions: With an Introduction to Sequency Theory,* Academic Press, New York, 1984.

2. Borut, A.O., Raczka, R.T., *Theory of Group Representations and Applications,* Varszawa, 1977.

3. Butzer, P.L., Nessel, R.J., *Fourier-Analysis and Approximation,* Vol. I, Birkhauser Verlag, Basel and Stuttgart, 1971.

4. Chrestenson, N.E., "A class of generalized functions", *Pacific J. Math.*, 5, 1955, 17-31.

5. Cooley, J.W., Tukey, C.W., "An algorithm for machine calculation of complex Fourier series", *Math. Computation*, Vol.19, 1965, 297-301.

6. Curtis, C.N., Rainer, I., *Representation Theory of Finite Groups and Associative Algebras,* Halsted, New York, London, 1962.

7. Fine, N.J., "On the Walsh functions", *Trans. Amer. Math. Soc.*, Vol. 65, No. 3, 1949, 374-414.

8. Hewitt, E., Ross, K.A., *Abstract Harmonic Analysis,* I,II, Springr-Verlag, Berlin, 1963, 1970.

9. Hurst, S.L., Miller, D.M., Muzio, J.C., *Spectral Techniques in Digital Logic,* Academic Press, Toronto, 1985.

10. Karpovsky, M.G., "Fast Fourier transforms on finite non-Abelian groups", *IEEE Trans. Computers*, Vol.C-26, 1028-1030, 1977.

11. Knapp, A.W., "Group representations and harmonic analysis from Euler to Langlands", Part 2, *Notices of AMS*, Vol. 43, No. 5, 537-549.

12. Moraga, C., "On a property of the Chrestenson spectrum", *IEE Proc.*, Vol. 129, Pt. E, 1982, 217-222.

13. Peter, F., Weyl, H., "Vollstandingkeit der primitiven Darstellungen einer geschlossen kontinuierlichen Gruppe", *Math. Ann.*, 97, 1927, 737-755.

14. Roziner, T.D., Karpovsky, M.G., Trachtenberg, L.A., "Fast Fourier transform over finite groups by multiprocessor systems", *IEEE Trans. Acoust., Speech, Signal Processing*, Vol. ASSP-38, No. 2, 226-240, 1990.

15. Stanković, R.S., "Matrix interpretation of fast Fourier transform on finite non-Abelian groups", *Res. Rept. in Appl. Math.*, YU ISSN 0353-6491, Ser. Fourier Analysis, Rept. No. 3, April 1990, 1-31, ISBN 86-81611-03-8.

16. Stanković, R.S., "Matrix interpretation of the fast Fourier transforms on finite non-Abelian groups", *Proc. Int. Conf. on Signal Processing, Beijing/90*, 22.-26.10.1990, Beijing, China, 1187-1190.

17. Stanković, R.S., Stojić, M.R., Bogdanović, M.R., *Fourier Representation of Signals*, Naučna knjiga, Beograd, 1988 (in Serbian).

18. Terras, A., *Fourier Analysis on Finite Groups and Applications*, Cambridge University Press, Cambridge, UK, 1999.

19. Vilenkin, N.Ya., "A class of orthonormal series", *Izv. Akad. Nauk SSSR, Ser. Mat.*, Vol. 11, 1947, 363-400.

20. Walsh, J.L., "A closed set of orthogonal functions", *Amer. J. Math.*, 45, 1923, 5-24.

3

Matrix Interpretation of the Fast Fourier Transform

There exist in each area of science some important concepts representing the corner stones of a whole theory and a corresponding practice, study of which, from different aspects, never goes out of interest. The algorithms for efficient computation of the discrete Fourier transform (DFT), generally known as the fast Fourier transform (FFT), form certainly such a concept in digital signal processing. Recall that the great practical application of the DFT and its importance in signal processing stems from the publication of the famous work by Cooley and Tukey in 1965 [10], though relatively recently interesting facts were discovered about the history of this algorithm [18].

Presently there is variety of algorithms in the quite voluminous literature on FFT, each of them suitable with respect to some a priori assumed criteria of optimality. These criteria are very different and range from the reduction of the time and memory resources needed for the computation to the use of some particular properties of the functions whose DFT is to be determined, or the use of the properties of spectral coefficients which should be calculated. Note for example, the real or pure imaginary functions, the symmetric functions, the functions with a lot of zero values, and similarly for spectral coefficients. For more detail see, for example, [1], [7],[31].

FFT algorithms are extended to be applicable to the calculation of the values of the generalized Fourier transform on finite Abelian groups [2], [8] including the DFT as a particular example.

Some other particular examples of this theory, as the Walsh or Vilenkin-Chrestenson transform, found also some important applications in different areas (see, for example, [3], [17], [22], [30]). Together with that, FFT algorithms were a base for the formulation of fast algorithms for the implementation of other discrete transforms on finite sets. Note as examples the discrete Haar transform, the slant transform, the discrete cosine transform (DCT), etc. More information about these algorithms can be found, for example, in [3] and the references therein. Moreover, the practical applicability of the discrete transforms mentioned above and many others is greatly supported by the existence of the fast algorithms for their implementation.

The matrix calculus appears to be a convenient way for representing discrete transforms and for dealing with them from theoretical, practical and educational point of view as well.

Among different discrete transforms, the Fourier transform on finite non-Abelian groups is recommended recently as the best choice in some particular applications [24], [25], [26], [54], [55]. As we already noted in Section 2.4, a fast algorithm for the implementation of this transform based on the classical Coley-Tukey FFT [31], is formulated in an analytical form in [23]. It seems that an earlier related result on this subject can be found in [11]. Matrix interpretation of FFT on finite non-Abelian groups has been given in [41].

The Fourier transform on finite non-Abelian groups can be studied in a unique setting with the Fourier transform on Abelian groups as well as the classical Fourier transform in the frame of abstract harmonic analysis on groups. However, in the case of Fourier transform on non-Abelian groups there are some important differences which must be appreciated at least in practical applications. According to this fact, our aim in this chapter is twofold. First, we consider a matrix representation of the fast Fourier transform on finite non-Abelian groups introduced in attempting to keep the entire analogy with the Abelian case as much as that is possible in the shape of the derived corresponding fast flow-graphs, and, then, we point out and discuss the main differences of this transform with respect to the fast Fourier transform on finite Abelian groups.

3.1   MATRIX INTERPRETATION OF FFT ON FINITE NON-ABELIAN GROUPS

To obtain a fast algorithm for the computation of the Fourier transform on finite non-Abelian groups we use the Good-Thomas method as in the case of the FFT on finite Abelian groups [15], [16], [52].

It is well known that the definition of the fast Fourier transform (FFT) on an Abelian group G (an algorithm for the efficient computation of the Fourier transform on G) is based upon the factorization of G into the equivalence classes relative to the subgroups of G. This group theoretical approach to the derivation of fast Fourier transform in the matrix notation can be interpreted as follows.

The disclosure of the FFT on a finite Abelian group G of the form (2.4) is based upon the factorization of the Fourier transform matrix into a product of sparse factors. Such a factorization is possible since the Fourier transform matrix on a given Abelian group G of the form (2.4) is representable as the Kronecker product of the Fourier transform matrices on its subgroups G_i. As is noted in Section 2.1, the transforms whose basic functions are generated as the Kronecker products of unitary irreducible representations of equal non-Abelian subgroups are also considered as the generalized Fourier, or short, Fourier transforms on groups [33].

Each of the factor matrices describes uniquely one step of the fast algorithm implementing the Fourier transform with respect to one particular coordinate x_i, $i = 1, \ldots, n$, determined by (2.5). In other words, the i-th step of the FFT can be considered as the restriction of the Fourier transform on the whole group G to the Fourier transform on its i-th subgroup G_i. It follows that the i-th factor matrix can be represented as the Kronecker product of the Fourier transformation matrix on G_i of order g_i at i-th position and the identity matrices of orders g_j, $j \in \{1, \ldots, n\} \backslash \{i\}$, at all other positions into that Kronecker product. We will extend the same approach to the non-Abelian groups by using the concepts of the generalized matrix multiplications.

The matrix $[\mathbf{R}]$ in the definition of the Fourier transform on finite non-Abelian groups is the matrix of unitary irreducible representations of G over P. Since G is representable in the form (2.4), the matrix $[\mathbf{R}]$ can be generated as the Kronecker product of $(K_i \times g_i)$ matrices $[\mathbf{R}_i]$ of unitary irreducible representations of subgroups G, $i \in \{1, \ldots, n\}$, i.e.,

$$[\mathbf{R}] = \bigotimes_{i=1}^{n} [\mathbf{R}_i].$$

Thanks to the well-known properties of the Kronecker product, the same applies to the matrix $[\mathbf{R}]^{-1}$, i.e., for this matrix holds

$$[\mathbf{R}]^{-1} = \bigotimes_{i=1}^{n} [\mathbf{R}_i]^{-1}.$$

This matrix can be further factorized into the elementwise Kronecker product of n sparse factors $[\mathbf{C}^i]$, $i \in \{1, \ldots, n\}$ as

$$[\mathbf{C}^i] = \bigotimes_{j=1}^{n} [\mathbf{S}_j^i], \quad i = 1, \ldots, n,$$

where

$$[\mathbf{S}_j^i] = \begin{cases} \mathbf{I}_{(g_j \times g_j)}, & j < i, \\ [\mathbf{R}_j]^{-1}, & j = i, \\ \mathbf{I}_{(K_j \times K_j)}, & j > i. \end{cases} \tag{3.1}$$

and $\mathbf{I}_{a \times a}$ is an $(a \times a)$ identity matrix.

Each matrix $[\mathbf{C}^i]$ describes uniquely one step of the fast Fourier transform performed in n steps. The algorithm is best represented by a flow-graph consisting of nodes connected with branches to which some weights are associated.

The matrix representation and the corresponding fast algorithm obtained in such a way is similar to the FFT on finite Abelian groups, but some important differences appear here.

As it is known, see, for example [22], the flow-graph of the i-th step of the FFT on a finite Abelian group G of order g has g input and g output nodes. The output nodes of the $(i-1)$-th step are the input nodes for the i-th step. The input nodes of the first step are the input nodes of the algorithm, and respectively, the output nodes of the n-th step are the output nodes of the algorithm except for the normalization with g^{-1}.

In the case of non-Abelian groups, the number of input and output nodes is different for the each step of the algorithm. Only the number of input nodes of the algorithm, i.e., the number of input nodes for the first step of the algorithm equals g. The number of input nodes of the i-th step is $g_1 g_2 \ldots g_{i-1} g_i K_{i+1} \ldots K_n$, while the number of output nodes is $g_1 g_2 \ldots g_{i-1} K_i K_{i+1} \ldots K_n$. Accordingly, the number of output nodes of the algorithm is equal to K.

The positions of non-zero elements of $[\mathbf{C}^i]$ determine which nodes will be connected. The weights associated to the branches are equal to the values of these elements. An important difference with respect to the FFT on Abelian groups is that in the case of non-Abelian groups the weights may be matrices, and therefore, according to the notation adopted in this book, will be denoted by bold letters. Denote by $k(i, j)$ the branch connecting the output node j with the input node i in the flow-graph of the k-th step of the FFT on a finite group. The weight $\mathbf{q}^k(i, j)$ associated to this branch is determined by $\mathbf{q}^k(i, j) = \mathbf{C}_{ij}^{n-k}$, where $\mathbf{C}_{ij}^{n-k}$ is the (i, j)-th element of the matrix $[\mathbf{C}^{n-k}]$. The branches for which the weight is equal to zero, i.e., the branches corresponding to zero elements of $[\mathbf{C}^{n-k}]$, do not appear in the flow-graph.

Now let us give a brief analysis of the complexity of the algorithm described here. The number of calculations is usually employed as a first approximation to the complexity of an algorithm.

Taking into account the g input nodes, the number of nodes in the flow-graph described, and hence, the number of basic operations $L(G)$ in the FFT on a finite non-Abelian group G based on this flow-graph, is equal to

$$L(G) = \sum_{i=1}^{n} a_i,$$

where $a_i = g_1 g_2 \ldots g_{i-1} K_i K_{i+1} \ldots K_n$.

Here by a basic operation in the i-th step of the algorithm we mean the operation given in a general form by

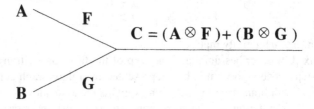

where $\mathbf{A}, \mathbf{B}$ are matrices of order $\Pi_{j=1}^{i-1} r_{w_j}$ while the weights $\mathbf{F}$ and $\mathbf{G}$ are matrices of order r_{w_j}. To obtain the Fourier coefficients $\mathbf{S}_f$ as they are defined by (2.14), it is needed to perform the normalization by g^{-1} after the calculation in the n-th step of algorithm is carried out.

Recall that in the case of Abelian groups, according to the definition of the unitary irreducible representations, all these matrices reduce to the numbers belonging to P, and hence, in that case the Kronecker product and matrix addition in P appearing in the basic operation defined here, reduce to the ordinary multiplication and addition in P.

Note that the flow-graph of the fast direct Fourier transform can be transformed into the flow-graph for the implementation of the inverse Fourier transform by a suitable mutual replacement of the input and output nodes, i.e., by considering the output nodes as the input nodes and vice versa. Clearly, the weights in this flow-graph are determined by the elements of the matrix $[\mathbf{R}]$ factorized as

$$[\mathbf{R}] = [\mathbf{D}^1] \overset{\circ}{\otimes} [\mathbf{D}^2] \overset{\circ}{\otimes} \ldots \overset{\circ}{\otimes} [\mathbf{D}^n],$$

where

$$[\mathbf{D}^i] = \bigotimes_{j=1}^{n} [\mathbf{E}_j^i], \quad i = 1, \ldots, n,$$

with

$$[\mathbf{E}_j^i] = \begin{cases} \mathbf{I}_{(K_j \times K_j)}, & j < i, \\ [\mathbf{R}_j]^{-1}, & j = i, \\ \mathbf{I}_{(g_j \times g_j)}, & j > i. \end{cases}$$

The main differences of the FFT on finite non-Abelian groups relative to the FFT on finite Abelian groups are summarized in Table 3.1.

3.2 ILLUSTRATIVE EXAMPLES

As it is usually the case with problems like those considered here, the algorithm is best explained by some examples.

Example 3.1 *Let $G_{2\times8}$ be a given group of order 16. The elements of this group will be denoted according to the convention adopted in this monograph by 0,1,...,15. The identity of the group is O, and the group operation is described in Table 3.2. All the unitary irreducible representations over the complex field C are given in Table 3.3. Note that in our notation, according to the definition of the inverse Fourier transform, the Table 3.3 defines the matrix $[\mathbf{R}]$ in (2.10).*

Note that the group $G_{2\times8}$ defined in this way can be considered as the direct product of the cyclic group C_2 of order 2 with modulo 2 addition as the group operation, and the quaternion group Q_2 of order 8.

The group representations of C_2 are $\begin{bmatrix} 1 & 1 \\ 1 & -1 \end{bmatrix}$, while these of the group Q_2 are in the left upper (5×8) quadrant in Table 3.3 of group representations. The

Table 3.1   Summary of differences between the FFT on finite Abelian and finite non-Abelian groups.

non-Abelian	Group G of order g	Abelian
non-Abelian		**Abelian**
dual object		dual object
Γ-the set of unitary irreducible representations		$\{\chi\}$-the set of group characters
Γ-does not have a group structure		$\{\chi\}$-has the structure of a multiplicative group
	Direct Fourier transform	
$\mathbf{S}_f(w) = g^{-1} r_w \sum_{x=0}^{g-1} f(x) \mathbf{R}_w(x^{-1}),$		$\widehat{f}(w) = g^{-1} \sum_{x=0}^{g-1} f(x) \overline{\chi}(w,x)$
	Number of spectral coefficients	
K		g
	Inverse Fourier transform	
$f(x) = \sum_{w=0}^{K-1} Tr(\mathbf{S}_f(w) \mathbf{R}_w(x))$		$f(x) = \sum_{w=0}^{g-1} \widehat{f}(x) \chi(w,x)$
	Fourier transformation matrix	
$[\mathbf{R}]^{-1}$		$[\overline{\chi}]$
	Order of the Fourier transformation matrix	
$K \times g$		$g \times g$
	Direct Fourier transform in matrix notation	
$[\mathbf{S}_f] = g^{-1}[\mathbf{R}]^{-1} \odot \mathbf{f}$		$\widehat{\mathbf{f}} = g^{-1}[\overline{\chi}]\mathbf{f}$
	Inverse Fourier transform in matrix notation	
$\mathbf{f} = [\mathbf{R}] \circ [\mathbf{S}_f]$		$\mathbf{f} = [\chi]\widehat{\mathbf{f}}$
	Number of steps of the fast algorithm	
n		n
	Number of input nodes	
1-st step g		g
i-th step $g_1 g_2 \ldots g_{i-1} g_i K_{i+1} \ldots K_n$		g
n-th step $g_1 K_2 \ldots K_n$		g
	Number of output nodes	
1-st step g		g
i-th step $g_1 g_2 \ldots g_{i-1} K_i K_{i+1} \ldots K_n$		g
n-th step K		g
	Weights in the flow-graph	
The values of unitary irreducible representations		The values of group characters
$\{r_w \mathbf{R}^{-1}(x)\}$		$\{\overline{\chi}(\cdot)\}$
$w \in \{0, \ldots, K-1\}, x \in \{0, \ldots, g-1\}$		$w, x \in \{0, 1, \ldots, g-1\}$
	Factor of normalization after n-th step	
g^{-1}		g^{-1}

cardinality of the dual object Γ *of* $G_{2\times 8}$ *is* $K = 10$, *and, according to (2.7) can be represented as* $K = K_1 K_2 = 2 \cdot 5$.

Notice that the ordering of the columns in Table 3.3 is the result of the usage of the Kronecker product to build the representations (compare the first 8 rows of the first five columns in Table 2.9).

The transformation matrix of the Fourier transform on $G_{2\times 8}$ *can be factorized as follows:*

$$[\mathbf{R}_{2\times 8}]^{-1} = [\mathbf{C}^1] \overset{\circ}{\otimes} [\mathbf{C}^2],$$

where

$$[\mathbf{C}^1] = \begin{bmatrix} 1 & 1 \\ 1 & -1 \end{bmatrix} \otimes \mathbf{I}_{5\times 5}, \tag{3.2}$$

$$[\mathbf{C}^2] = \mathbf{I}_{(2\times 2)} \otimes [\mathbf{Q}]^{-1}, \tag{3.3}$$

with $[\mathbf{Q}]^{-1}$ *as determined in Table 2.9 in Example 2.3.*

The flow-graph of the fast Fourier transform on G_{16} *corresponding to this factorization is shown in Figure 3.1. For simplicity the weights corresponding to the branches of this flow-graph are not indicated in this figure. As we noted above, these weights are determined by the values of the elements of the matrices (3.3) and (3.2) for the first and the second step of the flow-graph, respectively. For example,* $\mathbf{q}^1(5,4) = 2i\mathbf{D}/8$ *and* $\mathbf{q}^2(3,2) = 0$.

Note that in the example considered above the Fourier transform on the quaternion group Q_2 and C_2, is used as the basic module, and hence, is performed directly. In the examples below, we consider the Fourier transform on the symmetric group of permutations of order 3, S_3, in the same way. However, the fast algorithms for the computation of Fourier transform on these groups can be used here at least for the calculation of the coefficients corresponding to the unitary irreducible representations of order 1 and using the matrix operations for the calculation of the remaining coefficients. Of course, by a complete use of these fast algorithms, and therefore, by avoiding the matrix operations, the algorithms shown here can be easily translated into the ordinary-like FFT algorithms like those used in the case of Abelian groups. A discussion of this approach is given in [33].

The presented matrix interpretation of the FFT on finite non-Abelian groups requires the implementation of matrix operations in some branches of the algorithm. Therefore, the presented algorithms are efficient providing that the elements of the corresponding matrices are calculated simultaneously on some multiprocessor architectures. In other words, the matrix interpretation assumes that the constituent subgroups G_i of G are used as the basic modules, as in the case of Abelian groups, the Fourier transform on G is decomposed into n Fourier transforms on G_i, where it is implemented directly.

However, if the calculation of matrix-valued Fourier coefficients is decomposed into the calculation of their matrix elements, extending in that way the size of each

Table 3.2 Group operation of $G_{2\times 8}$.

	0	1	2	3	4	5	6	7	8	9	10	11	12	13	14	15
0	0	1	2	3	4	5	6	7	8	9	10	11	12	13	14	15
1	1	2	3	0	5	6	7	4	9	10	11	8	13	14	15	12
2	2	3	0	1	6	7	4	5	10	11	8	9	14	15	12	13
3	3	0	1	2	7	4	5	6	11	8	9	10	15	12	13	14
4	4	7	6	5	2	1	0	3	12	15	14	13	10	9	8	11
5	5	4	7	6	3	2	1	0	13	12	15	14	11	10	9	8
6	6	5	4	7	0	3	2	1	14	13	12	15	8	11	10	9
7	7	6	5	4	1	0	3	2	15	14	13	12	9	8	11	10
8	8	9	10	11	12	13	14	15	0	1	2	3	4	5	6	7
9	9	10	11	8	13	14	15	12	1	2	3	0	5	6	7	4
10	10	11	8	9	14	15	12	13	2	3	0	1	6	7	4	5
11	11	8	9	10	15	12	13	14	3	0	1	2	7	4	5	6
12	12	15	14	13	10	9	8	11	4	7	6	5	2	1	0	3
13	13	12	15	14	11	10	9	8	5	4	7	6	3	2	1	0
14	14	13	12	15	8	11	10	9	6	5	4	7	0	3	2	1
15	15	14	13	12	9	8	11	10	7	6	5	4	1	0	3	2

Table 3.3 Unitary irreducible representations of $G_{2\times 8}$ over C.

x	R_0	R_1	R_2	R_3	R_4	R_5	R_6	R_7	R_8	R_9
0	1	1	1	1	I	1	1	1	1	I
1	1	-1	1	-1	iA	1	-1	1	-1	iA
2	1	1	1	1	$-I$	1	1	1	1	$-I$
3	1	-1	1	-1	iB	1	-1	1	-1	iB
4	1	1	-1	-1	C	1	1	-1	-1	C
5	1	-1	-1	1	$-iD$	1	-1	-1	1	$-iD$
6	1	1	-1	-1	E	1	1	-1	-1	E
7	1	-1	-1	1	iD	1	-1	-1	1	iD
8	1	1	1	1	I	-1	-1	-1	-1	$-I$
9	1	-1	1	-1	iA	-1	1	-1	1	iB
10	1	1	1	1	$-I$	-1	-1	-1	-1	I
11	1	-1	1	-1	iB	-1	1	-1	1	iA
12	1	1	-1	-1	C	-1	-1	1	1	E
13	1	-1	-1	1	$-iD$	-1	1	1	-1	iD
14	1	1	-1	-1	E	-1	-1	1	1	C
15	1	-1	-1	1	iD	-1	1	1	-1	$-iD$

$$I = \begin{bmatrix} 1 & 0 \\ 0 & 1 \end{bmatrix} \qquad A = \begin{bmatrix} 1 & 0 \\ 0 & -1 \end{bmatrix} \qquad B = \begin{bmatrix} -1 & 0 \\ 0 & 1 \end{bmatrix}$$

$$C = \begin{bmatrix} 0 & -1 \\ 1 & 0 \end{bmatrix} \qquad D = \begin{bmatrix} 0 & 1 \\ 1 & 0 \end{bmatrix} \qquad E = \begin{bmatrix} 0 & 1 \\ -1 & 0 \end{bmatrix}$$

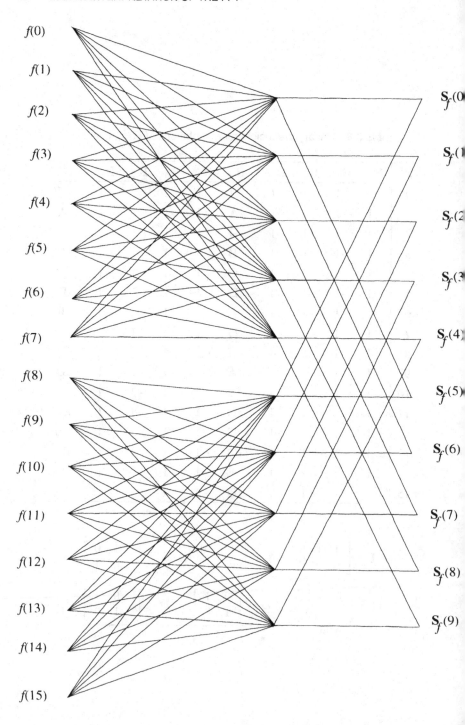

Fig. 3.1   Structure of the flow-graph of the FFT on the group $G_{2\times 8}$.

step of the FFT into g, the fast algorithms for the calculation of the FFT on some of the basic constituent subgroups can be applied by taking the advantage of some peculiar properties of group representations of the constituent subgroups. For example, the calculation of Fourier transform on the quaternion group Q_2 can be done without multiplication except the multiplication by the imaginary unit i, see Example 2.3. The statement will be illustrated by the following example.

Example 3.2 *Let G_{32} be the direct product of two cyclic groups C_2 of order 2 with modulo 2 addition as the group operation and the quaternion group Q_2 of order 8, i.e., $G_{32} = C_2 \times C_2 \times Q_2$. The Fourier transformation matrix on this group over the complex field can be factorized as*

$$[\mathbf{R}]^{-1} = [\mathbf{C}^1] \overset{\circ}{\otimes} [\mathbf{C}^2] \overset{\circ}{\otimes} [\mathbf{C}^3],$$

where

$$[\mathbf{C}^1] = \begin{bmatrix} 1 & 1 \\ 1 & -1 \end{bmatrix} \otimes \mathbf{I}_{2\times 2} \otimes \mathbf{I}_{5\times 5},$$

$$[\mathbf{C}^2] = \mathbf{I}_{2\times 2} \otimes \begin{bmatrix} 1 & 1 \\ 1 & -1 \end{bmatrix} \otimes \mathbf{I}_{5\times 5},$$

$$[\mathbf{C}^3] = \mathbf{I}_{2\times 2} \otimes \mathbf{I}_{2\times 2} \otimes [\mathbf{Q}_2]^{-1},$$

with the matrix $[\mathbf{Q}_2]$ described in Example 2.3.

The structure of the flow-graph of the FFT on G_{32} derived according to this factorization is given in Figure 3.2. For simplicity, the weighting coefficients are not shown at the edges of the graph, and these values are determined by the values of the corresponding matrix elements in the factorization of transform matrices.

In this algorithm the Fourier transforms on the cyclic groups C_2 and the quaternion group Q_2 are used as the basic modules, and hence they are performed directly. The C_2 is the simplest possible case, but the Fourier transform on Q_2 can be carried out by using the corresponding fast algorithm. If we do not like to work with matrix operations in a FFT flow-graph, and if we use the fast algorithm on Q_2, the algorithm given in Figure 3.2 can be translated easily into an ordinary FFT like those used in the case of Abelian groups. In that order note that the first four rows of $[Q]^{-1}$ are identical to some particular Walsh functions on the finite group of order 8, and, hence, some of the values representing the output from the first step of our algorithm can be calculated by using a part of the flow-graph of the corresponding fast Walsh transform. The remaining output values from the first step correspond to the group representation of order 2, and therefore they are (2×2) matrices. The elements of these matrices will be calculated independently and for efficiency we will use the fact that each of the matrices $\mathbf{I}$, $\mathbf{A}$, $\mathbf{B}$, $\mathbf{C}$, $\mathbf{D}$, $\mathbf{E}$ contains two zero elements. In this way we derive the algorithm for the calculation of Fourier transform on G_{32} shown in Figure 3.3. This algorithm can be compared with the corresponding algorithm presented in [33]

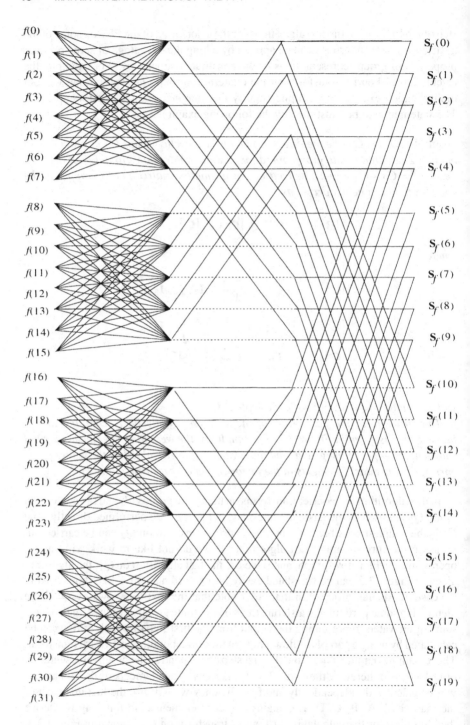

Fig. 3.2 Structure of the flow-graph for FFT on the group G_{32}.

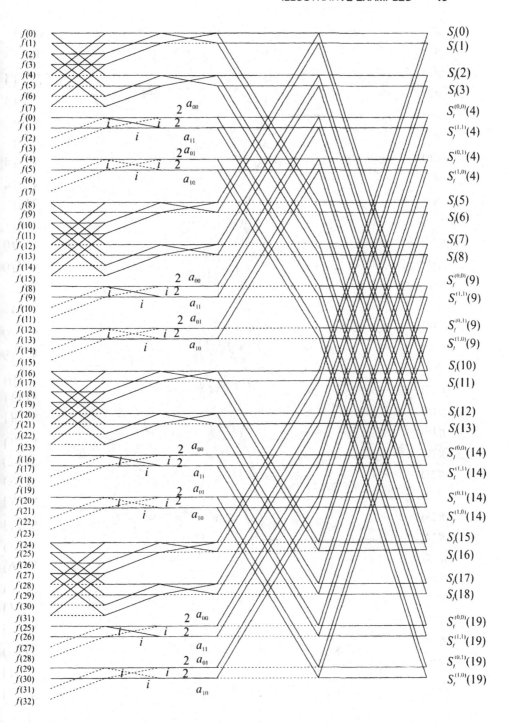

Fig. 3.3 Structure of the flow-graph for FFT on the group G_{32} through a part of fast Walsh transform.

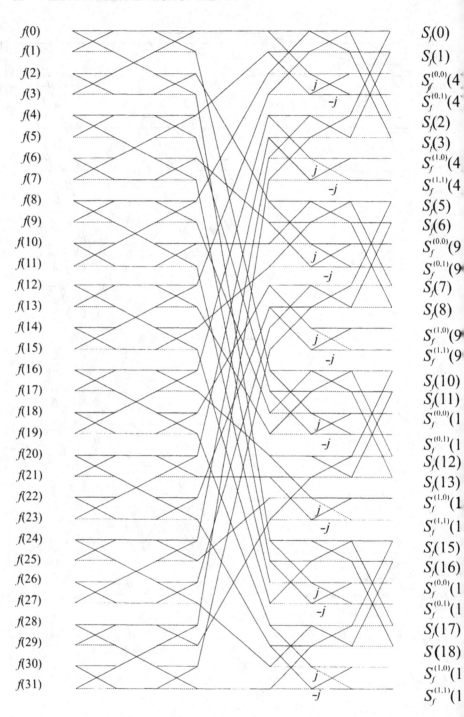

Fig. 3.4 Structure of the flow-graph for FFT on the group G_{32} using FFT on Q_2.

shown in Figure 3.4. In this algorithm a different fast algorithm for the calculation of the Fourier transform on the quaternion group described in Example 2.3 is used.

Example 3.3 *Let $S_3 = (0, (132), (123), (12), (13), (23), \circ)$ be the symmetric group of permutations of order 3 defined in Example 2.1. Let $G_{6\times 6} = S_3 \times S_3$ be the direct product of S_3 with itself, i.e., $G_{6\times 6}$ consists of pairs $(h_1, h_2) = g \in G_{6\times 6}$, where $h_1, h_2 \in S_3$. The group operation of $G_{6\times 6}$ is specified as follows: for $(h_1, h_2) = g \in G_{6\times 6}$, and $(h'_1, h'_2) = g' \in G_{6\times 6}$, we have $(h_1 \overset{\circ}{s} h'_1, h_2 \overset{\circ}{s} h'_1) = g \circ g' \in G_{6\times 6}$.*

The group operation table is large and we do not write it explicitly. It can be easily derived from the group operation table for S_3. The unitary irreducible representations of $G_{6\times 6}$ over the Galois field GF(11) are given in Table 3.4.

The Fourier transform matrix on $G_{6\times 6}$ can be factorized as

$$[\mathbf{R}_{6\times 6}]^{-1} = [\mathbf{C}^1] \overset{\circ}{\otimes} [\mathbf{C}^2],$$

where

$$[\mathbf{C}^1] = [\mathbf{S}_3] \otimes [\mathbf{I}_{3\times 3}],$$
$$[\mathbf{C}^2] = [\mathbf{I}_{6\times 6}] \otimes [\mathbf{S}_3],$$

with

$$[\mathbf{S}_3]^{-1} = 2 \begin{bmatrix} 1 & 1 & 1 & 1 & 1 & 1 \\ 1 & 1 & 1 & 10 & 10 & 10 \\ 2\mathbf{I} & 2\mathbf{B} & 2\mathbf{A} & 2\mathbf{C} & 2\mathbf{D} & 2\mathbf{E} \end{bmatrix},$$

with the notation as in Table 2.5 since $\mathbf{A} = \mathbf{B}^T$, and $\mathbf{C}, \mathbf{D}, \mathbf{E}$ are symmetric matrices.

The flow-graph of the fast Fourier transform on $G_{6\times 6}$ based on this factorization is shown in Figure 3.5.

Example 3.4 *Let $G_{3\times 6}$ be the direct product of the group $Z_3 = (0, 1, 2, \overset{\circ}{3})$ of integers less than 3 with modulo 3 addition as the group operation, and the symmetric group of permutations of order 3, S_3 described in Example 2.1. The group Z_3 is an Abelian group, and therefore, their representations are given by the matrix of characters $[\chi]$ as follows:*

$$[\chi] = \begin{bmatrix} 1 & 1 & 1 \\ 1 & e_1 & e_2 \\ 1 & e_2 & e_1 \end{bmatrix},$$

where $e_1 = -2^{-1}(1 - i\sqrt{3}), e_2 = -2^{-1}(1 + i\sqrt{3})$. The unitary irreducible representations of S_3 over C are given in Table 2.2. Hence, $G_{3\times 6}$ consists of pairs $(h_1, h_2) = g \in G_{3\times 6}$ where $h_1 \in Z_3$ and $h_2 \in S_3$. The group operation $\circ$ of $G_{3\times 6}$ is specified as follows: for $(h_1, h_2) = g \in G_{3\times 6}$ and $(h'_1, h'_2) = g' \in G_{3\times 6}$ we have $(h_1 \overset{\circ}{3} h'_1, h_2 \overset{\circ}{3} h'_2) = g \circ g' \in G_{3\times 6}$. The group table of $G_{3\times 6}$ is given in Table 3.5, while the unitary irreducible representations of $G_{3\times 6}$ are given in Table 3.6.

Table 3.4 The unitary irreducible representations of $G_{6 \times 6}$ over $GF(11)$.

x	$\mathbf{R}_0$	$\mathbf{R}_1$	$\mathbf{R}_2$	$\mathbf{R}_3$	$\mathbf{R}_4$	$\mathbf{R}_5$	$\mathbf{R}_6$	$\mathbf{R}_7$	$\mathbf{R}_8$
0	1	1	I	1	1	I	I	I	$I \otimes I$
1	1	1	A	1	1	A	I	I	$I \otimes A$
2	1	1	B	1	1	B	I	I	$I \otimes B$
3	1	10	C	1	10	C	I	10I	$I \otimes C$
4	1	10	D	1	10	D	I	10I	$I \otimes D$
5	1	10	E	1	10	E	I	10I	$I \otimes E$
6	1	1	I	1	1	I	A	A	$A \otimes I$
7	1	1	A	1	1	A	A	A	$A \otimes A$
8	1	1	B	1	1	B	A	A	$A \otimes B$
9	1	10	C	1	10	C	A	10A	$A \otimes C$
10	1	10	D	1	10	D	A	10A	$A \otimes D$
11	1	10	E	1	10	E	A	10A	$A \otimes E$
12	1	1	I	1	1	I	B	B	$B \otimes I$
13	1	1	A	1	1	A	B	B	$B \otimes A$
14	1	1	B	1	1	B	B	B	$B \otimes B$
15	1	10	C	1	10	C	B	10B	$B \otimes C$
16	1	10	D	1	10	D	B	10B	$B \otimes D$
17	1	10	E	1	10	E	B	10B	$B \otimes E$
18	1	1	I	10	10	10I	C	C	$C \otimes I$
19	1	1	A	10	10	10A	C	C	$C \otimes A$
20	1	1	B	10	10	10B	C	C	$C \otimes B$
21	1	10	C	10	1	10C	C	10C	$C \otimes C$
22	1	10	D	10	1	10D	C	10C	$C \otimes D$
23	1	10	E	10	1	10E	C	10C	$C \otimes E$
24	1	1	I	10	10	10I	D	D	$D \otimes I$
25	1	1	A	10	10	10A	D	D	$D \otimes A$
26	1	1	B	10	10	10B	D	D	$D \otimes B$
27	1	10	C	10	1	10C	D	10D	$D \otimes C$
28	1	10	D	10	1	10D	D	10D	$D \otimes D$
29	1	10	E	10	1	10E	D	10D	$D \otimes E$
30	1	1	I	10	10	10I	E	E	$E \otimes I$
31	1	1	A	10	10	10A	E	E	$E \otimes A$
32	1	1	B	10	10	10B	E	E	$E \otimes B$
33	1	10	C	10	1	10C	E	10E	$E \otimes C$
34	1	10	D	10	1	10D	E	10E	$E \otimes D$
35	1	10	E	10	1	10E	E	10E	$E \otimes E$

I, A, B, C, D, E as in Table 2.5.

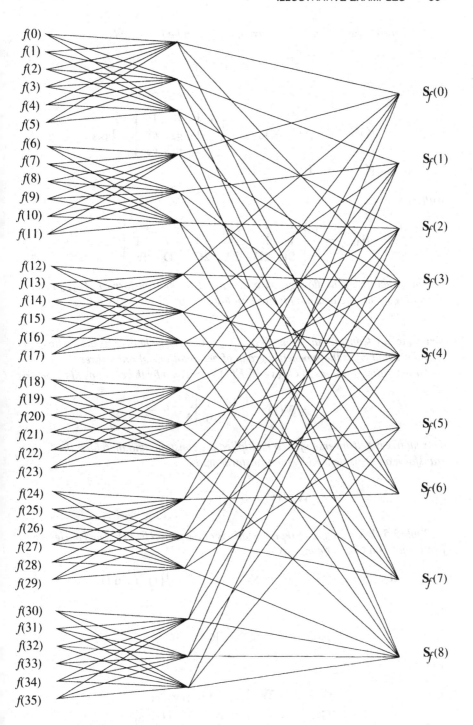

Fig. 3.5 Structure of the flow-graph for FFT on the group $G_{6\times6}$.

The Fourier transformation matrix on $G_{3\times 6}$ can be factorized as

$$[\mathbf{R}]^{-1} = [\mathbf{C}^1]\,[\mathbf{C}^2],$$

where

$$[\mathbf{C}^1] = [\chi]^{-1} \otimes \mathbf{I}_{3\times 3} = \frac{1}{3}\begin{bmatrix} 1 & 1 & 1 \\ 1 & e_2 & e_1 \\ 1 & e_1 & e_2 \end{bmatrix} \otimes \mathbf{I}_{3\times 3},$$

$$[\mathbf{C}^2] = \mathbf{I}_{3\times 3} \otimes [\mathbf{S}_3]^{-1},$$

with

$$[\mathbf{S}_3]^{-1} = \frac{1}{6}\begin{bmatrix} 1 & 1 & 1 & 1 & 1 & 1 \\ 1 & 1 & 1 & -1 & -1 & -1 \\ 2\mathbf{I} & 2\mathbf{B} & 2\mathbf{A} & 2\mathbf{C} & 2\mathbf{D} & 2\mathbf{E} \end{bmatrix}.$$

The structure of the flow-graph of the fast Fourier transform on $G_{3\times 6}$ derived according to this factorization is given in Figure 3.6.

Example 3.5 *Consider the group $G_{24} = C_2 \times C_2 \times S_3$, where C_2 is the cyclic group of order 2 and S_3 is the symmetric group of permutations of order three.*

Group representations of C_2 over $GF(11)$ are given by the columns of the matrix

$$\mathbf{W}(1) = \begin{bmatrix} 1 & 1 \\ 1 & 10 \end{bmatrix}.$$

This matrix is self-inverse up to a multiplicative constant, and therefore, the Fourier transform on C_2 is defined by the transform matrix

$$\mathbf{W}^{-1}(1) = 6\begin{bmatrix} 1 & 1 \\ 1 & 10 \end{bmatrix}.$$

Table 2.5 shows the group representations of S_3 over $GF(11)$. The Fourier transform matrix on G_{24} is defined by

$$[\mathbf{R}_{24}]^{-1}(3) = \left(\mathbf{W}^{-1}(1) \otimes \mathbf{W}^{-1}(1) \otimes [\mathbf{S}_3]^{-1}(1)\right) \bmod (11).$$

$$[\mathbf{R}_{24}]^{-1}(3) = [\mathbf{C}^1] \overset{\circ}{\otimes} [\mathbf{C}^2] \overset{\circ}{\otimes} [\mathbf{C}^3],$$

where

$$[\mathbf{C}^1] = \mathbf{W}^{-1}(1) \otimes \mathbf{I}_{(2\times 2)} \otimes \mathbf{I}_{(3\times 3)},$$
$$[\mathbf{C}^2] = \mathbf{I}_{(2\times 2)} \otimes \mathbf{W}^{-1}(1) \otimes \mathbf{I}_{(3\times 3)},$$
$$[\mathbf{C}^3] = \mathbf{I}_{(2\times 2)} \otimes \mathbf{I}_{(2\times 2)} \otimes [\mathbf{S}_3]^{-1}(1).$$

Table 3.5 The group operation of $G_{3 \times 6}$.

	0	1	2	3	4	5	6	7	8	9	10	11	12	13	14	15	16	17
0	0	1	2	3	4	5	6	7	8	9	10	11	12	13	14	15	16	17
1	1	2	0	5	3	4	7	8	6	11	9	10	13	14	12	17	15	16
2	2	0	1	4	5	3	8	6	7	10	11	9	14	12	13	16	17	15
3	3	4	5	0	1	2	9	10	11	6	7	8	15	16	17	12	13	14
4	4	5	3	2	0	1	10	11	9	8	6	7	16	17	15	14	12	13
5	5	3	4	1	2	0	11	9	10	7	8	6	17	15	16	13	14	12
6	6	7	8	9	10	11	12	13	14	15	16	17	0	1	2	3	4	5
7	7	8	6	11	9	10	13	14	12	17	15	16	1	2	0	5	3	4
8	8	6	7	10	11	9	14	12	13	16	17	15	2	0	1	4	5	3
9	9	10	11	6	7	8	15	16	17	12	13	14	3	4	5	0	1	2
10	10	11	9	8	6	7	16	17	15	14	12	13	4	5	3	2	0	1
11	11	9	10	7	8	6	17	15	16	13	14	12	5	3	4	1	2	0
12	12	13	14	15	16	17	0	1	2	3	4	5	6	7	8	9	10	11
13	13	14	12	17	15	16	1	2	0	5	3	4	7	8	6	11	9	10
14	14	12	13	16	17	15	2	0	1	4	5	3	8	6	7	10	11	9
15	15	16	17	12	13	14	3	4	5	0	1	2	9	10	11	6	7	8
16	16	17	15	14	12	13	4	5	3	2	0	1	10	11	9	8	6	7
17	17	15	16	13	14	12	5	3	4	1	2	0	11	9	10	7	8	6

Table 3.6 The unitary irreducible representations of $G_{3\times6}$ over C.

x	R_0	R_1	R_2	R_3	R_4	R_5	R_6	R_7	R_8
0	1	1	**I**	1	1	**I**	1	1	**I**
1	1	1	**A**	1	1	**A**	1	1	**A**
2	1	1	**B**	1	1	**B**	1	1	**B**
3	1	-1	**C**	1	-1	**C**	1	-1	**C**
4	1	-1	**D**	1	-1	**D**	1	-1	**D**
5	1	-1	**E**	1	-1	**E**	1	-1	**E**
6	1	1	**I**	e_1	e_1	$e_1\mathbf{I}$	e_2	e_2	$e_2\mathbf{I}$
7	1	1	**A**	e_1	e_1	$e_1\mathbf{A}$	e_2	e_2	$e_2\mathbf{A}$
8	1	1	**B**	e_1	e_1	$e_1\mathbf{B}$	e_2	e_2	$e_2\mathbf{B}$
9	1	-1	**C**	e_1	$-e_1$	$e_1\mathbf{C}$	e_2	$-e_2$	$e_2\mathbf{C}$
10	1	-1	**D**	e_1	$-e_1$	$e_1\mathbf{D}$	e_2	$-e_2$	$e_2\mathbf{D}$
11	1	-1	**E**	e_1	$-e_1$	$e_1\mathbf{E}$	e_2	$-e_2$	$e_2\mathbf{E}$
12	1	1	**I**	e_2	e_2	$e_2\mathbf{I}$	e_1	e_1	$e_1\mathbf{I}$
13	1	1	**A**	e_2	e_2	$e_2\mathbf{A}$	e_1	e_1	$e_1\mathbf{A}$
14	1	1	**B**	e_2	e_2	$e_2\mathbf{B}$	e_1	e_1	$e_1\mathbf{B}$
15	1	-1	**C**	e_2	$-e_2$	$e_2\mathbf{C}$	e_1	$-e_1$	$e_1\mathbf{C}$
16	1	-1	**D**	e_2	$-e_2$	$e_2\mathbf{D}$	e_1	$-e_1$	$e_1\mathbf{D}$
17	1	-1	**E**	e_2	$-e_2$	$e_2\mathbf{E}$	e_1	$-e_1$	$e_1\mathbf{E}$

The notation as in Table 2.2.

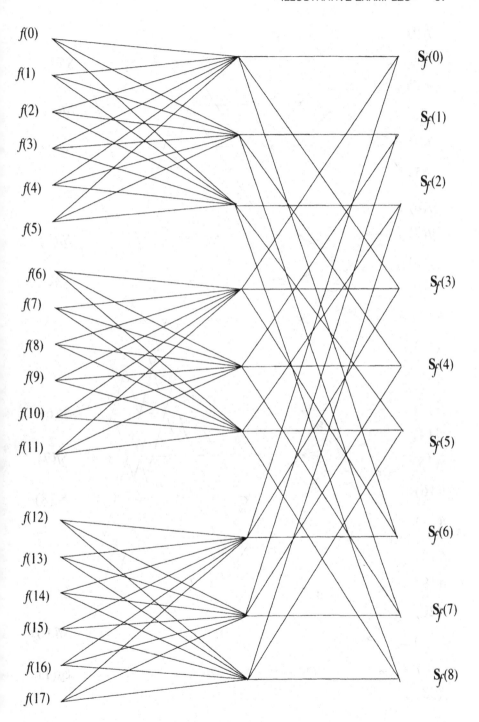

Fig. 3.6 Structure of the flow-graph for FFT on the group $G_{3 \times 6}$.

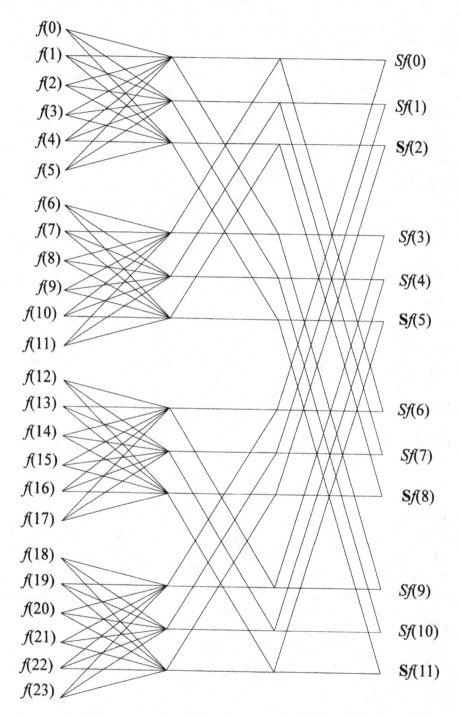

Fig. 3.7 Structure of the flow-graph for FFT on G_{24}.

Figure 3.7 shows the flow-graph of FFT on G_{24} derived from this factorization of $[\mathbf{R}_{24}]^{-1}(3)$. *In this flow-graph, the weights at the edges are elements of* $[\mathbf{S}_3](1)$, *and of* $\mathbf{W}(1)$.

As described above, in Example 3.2, with FFT the calculation of one-dimensional Fourier transform on G of the form (2.4) is transferred into successive calculation of n Fourier transforms on the constituent subgroups G_i of G. Each block in the related flow-graph corresponds to a subgroup G_i. In these blocks, calculation of the Fourier transform on G_i is performed directly by definition of the transform. Some further reduction of computations can be achieved if it is possible to develop some FFT on the constituent subgroups [54].

The Fourier transform matrix on S_3 written in terms of functions $\mathbf{R}_w^{(i,j)}(x)$ shown in Table 2.7, is given by

$$\mathbf{S}_3^{-1}(1) = 2 \begin{bmatrix} 1 & 1 & 1 & 1 & 1 & 1 \\ 1 & 1 & 1 & 10 & 10 & 10 \\ 2 & 10 & 10 & 2 & 10 & 10 \\ 0 & 5 & 6 & 0 & 5 & 6 \\ 0 & 6 & 5 & 0 & 5 & 6 \\ 2 & 10 & 10 & 9 & 1 & 1 \end{bmatrix}$$

This matrix can be factorized as follows:

$$\mathbf{S}_3(1) = 2 \begin{bmatrix} 1 & 0 & 0 & 10 & 0 & 0 \\ 1 & 0 & 0 & 1 & 0 & 0 \\ 0 & 1 & 0 & 0 & 10 & 0 \\ 0 & 0 & 1 & 0 & 0 & 10 \\ 0 & 0 & 10 & 0 & 0 & 10 \\ 0 & 1 & 0 & 0 & 1 & 0 \end{bmatrix} \begin{bmatrix} 1 & 1 & 1 & 0 & 0 & 0 \\ 2 & 10 & 10 & 0 & 0 & 0 \\ 0 & 5 & 6 & 0 & 0 & 0 \\ 0 & 0 & 0 & 10 & 10 & 10 \\ 0 & 0 & 0 & 9 & 1 & 1 \\ 0 & 0 & 0 & 0 & 6 & 5 \end{bmatrix} \quad \text{mod (11)}.$$

This factorization produces the FFT on S_3 over $GF(11)$. Figure 3.8 shows the flow-graph of this FFT on S_3. In this figure, the ordering of elements of $\mathbf{S}_f(2)$ as $S_f(2)^{(0,0)}$, $S_f(2)^{(1,1)}$, $S_f(2)^{(1,0)}$, $S_f(2)^{(0,1)}$, makes the graph symmetric.

Figure 3.9 shows the FFT on G_{24} derived by using FFT on S_3. In this flow-graph all the weights are numbers. The output is the vector of number-valued Fourier coefficients $S_f(0)$, $S_f(1)$, and elements of the matrix-valued Fourier coefficients $S_f^{(i,j)}(2)$, $i, j = 0, 1$, for f.

3.3 COMPLEXITY OF THE FFT

In this section, we compare efficiency of the FFT on finite dyadic groups and the quaternion groups. Rationales for this comparison can be found in the following considerations.

In some areas as, for example, the switching theory and logic design [36], the application of spectral methods, although theoretically proven very efficient [1], [3],

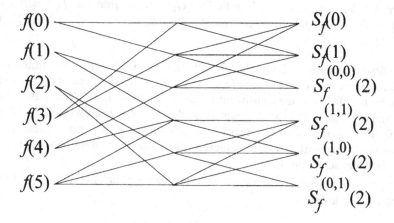

Fig. 3.8   Structure of the flow-graph for FFT on S_3.

[20], [19], [22], [56], is rather restricted in practice for the exponential complexity of the FFT and related algorithms. In these areas, it is often required to calculate with switching functions of a large number of variables, for example, $n > 20$ or more.

In spectral methods, the switching functions, being dependent on two-valued variables, are naturally considered as functions on the dyadic groups. At the same time, being two-valued functions, the set of all switching functions of n-variables is the dyadic group of order $C_2^{2^n}$ under the componentwise EXOR performed over truth-vectors.

These basic features determine complexity of representations, manipulations and calculations with switching functions. If n is large, these features become a restrictive factor for many methods and techniques. For example, determination of coefficients in spectral transform representations, as for example, the Reed-Muller expressions and related Kronecker expressions [35], [47], by FFT-like algorithms [4], [5], expresses exponential complexity of $O(2^n)$ and $O(n2^n)$ in terms of space and time. The same applies to arithmetic, Walsh, and other related representations of switching functions [51]. Thus, if n is large, for example $n > 30$, such algorithms are hardly applicable on most of standard computer architectures.

In [54], the following question is asked: "The ultimate purpose must be to find out whether this group (the quaternion Q_2) may be as significant for logic synthesis as it seems to be for filtering and other signal processing tasks".

The following considerations, based on [45] and [46], are an attempt towards approaching an answer to this question. The switching functions are considered as elements of vector spaces on the dyadic groups and the quaternion groups over C.

We assume that in an n-variable switching function f, each triplet of variables x_i, x_j, x_k, where each variable takes values in C_2, is replaced by a variable X_r on

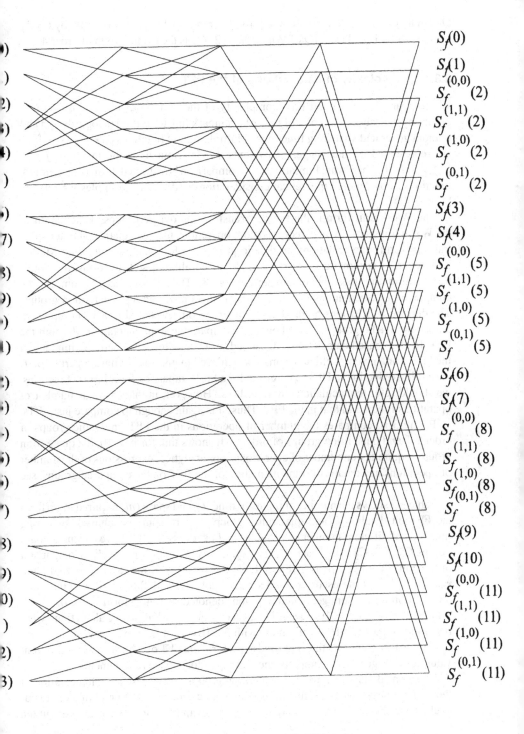

Fig. 3.9 Structure of the flow-graph for FFT on G_{24} with FFT on S_3.

Q_2. In that way, the domain group C_2^n for f is replaced by Q_2^k, for $n = 3k$, by $C_2 Q_2^k$ for $n = 3k + 1$, and by $C_4 Q_2^k$ for $n = 3k + 2$. C_4 is the cyclic group of order 4.

3.3.1 Complexity of calculations of the FFT

We compare the space and time complexity of the FFT on dyadic and quaternion groups by the way of examples of *mcnc* benchmark functions used in logic design and random generated switching functions. Multiple-output functions $f_0 * f_1 * \cdots * f_{q-1}$ are represented by the integer equivalents f_Z determined by the mapping $f_Z = \sum_{i=0}^{q-1} 2^i f_i$. For hardware limitations, the comparison is restricted to small benchmark functions. For detailed theoretical considerations and further examples we refer to [45].

The Walsh transform is the Fourier transform on dyadic groups. Calculation of the Walsh transform by FFT requires more operations than calculation of other spectral transform representations on dyadic groups, as for example, the Reed-Muller, Kronecker, or arithmetic expressions [14], since unlike these transforms, the Walsh transform matrix does not contain zero elements. Thus, it is enough to compare the complexity of calculations of Fourier transforms on dyadic and quaternion groups. The derived conclusions holds even stronger for other spectral transform representations on dyadic groups. It is enough to compare the implementations through the FFT. In DDs based calculations, discussed in what follows, we actually implement the FFT. For a given f, efficiency may be achieved thanks to peculiar properties of f, permitting reduction in DDs. Therefore, for a fair comparison for arbitrary functions, we should refer to calculations through DTs. However, in that case, the number of operations is equal to that in the FFT. Thus, it is again enough to compare just FFTs.

Figure 3.10 compares the number of operations in the FFT on dyadic groups of order 2^{3r} and quaternion groups of order r. It shows that for $n > 10$, the quaternion groups require fewer number of operations. For analytical expressions of the number of operations and details of implementation of the FFT on the compared groups, see [45].

Table 3.7 shows the number of inputs n, time (t), and space (m) required to perform the FFT on dyadic (d) and quaternion groups (q) for some benchmark switching functions of different number of variables n. For $n = 8, 9, 10, 14$, we use the domain groups C_2^8, C_2^9, C_2^{10}, C_2^{14}, and $C_4 Q_2^2$, Q_2^3, $C_2 Q_2^3$, $C_4 Q_2^4$, respectively. Comparison of the considered groups is given in Table 3.8 in terms of the ratio $r_t = t_d/t_q$ and $r_m = m_d/m_q$ of the used time and space. Table 3.9 and Table 3.10 give the same information for random generated functions denoted by fun(n) for $n = 9, 10, 12, 14$. Time unit is $1msec = 10^{-3}sec$. The space is given in KBytes. Calculations were performed on a 133MHz Pentium PC with 32MBytes of main memory.

Figure 3.11 compares time required to perform the FFT on C_2^{3r} and Q_2^r for different values of r. Figure 3.12 compares the corresponding space requirements.

This consideration shows that quaternion groups permit faster implementation of the FFT for large switching functions than dyadic groups, at the price of a comparatively small increase of the required space. For example, for $n > 10$ for at about ten

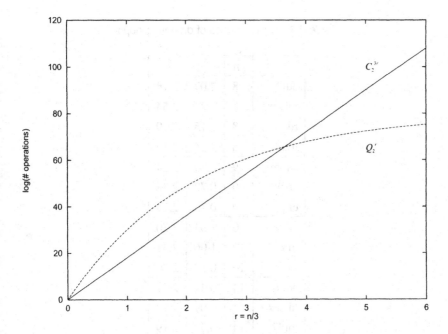

Fig. 3.10 Number of operations in FFT.

Table 3.7 Complexity of the FFT.

f	n	t_d	m_d	t_q	m_q
adr4	8	60.90	46.39	125.80	100.27
misex1	8	33.20	35.14	73.70	89.17
rd84	8	116.10	46.39	133.90	101.39
mul4	8	63.40	45.27	129.70	99.29
9sym	9	379.80	86.39	324.60	193.30
apex4	9	464.00	86.39	336.00	196.96
clip	9	466.00	86.39	339.00	197.38
adr5	10	759.00	166.39	787.00	381.89
mul5	10	788.00	164.14	833.00	379.93
sao2	10	1531.00	166.39	763.00	375.99
adr7	14	233360.00	1926.39	78360.00	5334.39
misex3	14	801480.00	1926.39	75060.00	5331.19
mul7	14	123200.00	1618.28	65320.00	5165.25

Table 3.8 Comparisons of domain groups.

f	n	r_t	r_m
adr4	8	2.07	2.16
misex1	8	2.22	2.54
rd84	8	1.15	2.19
mul4	8	2.04	2.19
9sym	9	0.85	2.24
apex4	9	0.72	2.28
clip	9	0.73	2.28
adr5	10	1.04	2.30
mul5	10	1.06	2.31
sao2	10	0.50	2.26
adr7	14	0.34	2.77
misex3	14	0.09	2.77
mul7	14	0.53	3.19

Table 3.9 FFT for random functions.

f	t_d	m_d	t_q	m_q
fun(9)	536.00	86.39	352.00	198.64
fun(10)	2048.00	166.39	916.00	390.89
fun(12)	39620.00	486.39	7110.00	1356.32
fun(14)	949000.00	1930.30	94280.00	5406.07

Table 3.10 Comparison of the FFT for random functions.

f	r_t	r_m
fun(9)	0.66	2.30
fun(10)	0.45	2.35
fun(12)	0.18	2.79
fun(14)	0.10	2.80

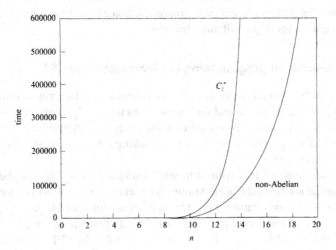

Fig. 3.11 Time requirements.

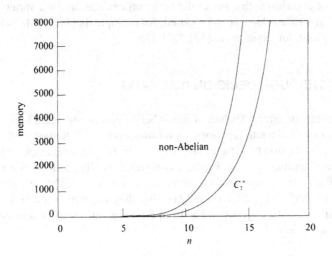

Fig. 3.12 Memory requirements.

times faster implementation of the FFT, three times more space is required. When the number of variables n grows, the advantages of quaternion groups increase.

It follows that, compared to dyadic groups, the quaternion groups have advantages as domain groups for large switching functions.

3.3.2 Remarks on programming implementation of FFT

It should be noticed that besides the number of operations, the programming implementation of a fast algorithm and the hardware used strongly affects its efficiency in terms of time. This aspect also requires some analysis in determination of a transform suitable for a particular application and taking into account also the hardware provided.

For a given spectral transform different factorization formulas can be derived. These formulas are derived from the same given set of factorization rules for different specification of various parameters. The number of different formulas is large and dependent on the transform size. For example, as pointed out in [38], there are possible 258400 and 1.8×1013 different formulas for the FFT of the size 25 and 26, respectively. Several different factorizations may require the same number of arithmetic operations, however, the order of implementation of the operations may considerably influence the calculation time extending it from 2 to 10 times. In this respect, a lot of research work has been done in searching for the best implementation formulas for the intended hardware. For instance, methods for determination of the most efficient fast Walsh transform with respect to the implementation time has been considered in [21], [32]. Extensions and generalizations of these approaches to the derivation of algorithms that select the best implementation for a variety of signal processing algorithms have been proposed, for example, in [12], [38]. Related work for FFT is given, for example, in [13], [27], [28].

3.4 FFT THROUGH DECISION DIAGRAMS

FFT algorithms on either Abelian or non-Abelian groups are based upon the vector representations of discrete functions. It follows from the definition of the FFT and their matrix description that space complexity of the FFT on a decomposable group G of order g approximates $O(g)$. The time complexity is $O(ng)$. Thus, the application of the FFT is restricted to groups of relatively small orders. This restriction can be overcome by performing FFT on the decision diagram representations of discrete functions in the same way as that has been done for various discrete transforms on finite Abelian groups.

3.4.1 Decision diagrams

A decision diagram (DD) is a data structure that can compactly represent discrete functions. It is derived from a Decision tree (DT) by merging decision routes that lead to same final value. In this way they often require much less space that other representations.

In this book, we use Multi-terminal decision diagrams (MTDDs) [29] to represent discrete functions on finite not necessarily Abelian groups. MTDDs are a generalization of Multi-terminal binary DDs (MTBDDs) [9]. MTBDDs were defined as a generalization of Binary DDs (BDDs) [6] by permitting complex numbers as the values of constant nodes. Thus, they can be used to represent complex-valued functions on dyadic groups. Multiple-place decision diagrams (MDDs) are a generalization of MTBDDs from dyadic groups to groups which are direct product of cyclic groups of order p, C_p, [29], [39], [49], [50]. They are derived by allowing p outgoing edges for each node in the DD. MTDDs are a further generalization of MDDs derived by allowing that nodes at different levels in the MTDD may have different number of outgoing edges.

MTDDs are derived by the reduction of Multi-terminal decision trees (MTDTs), which can be introduced through the following considerations.

A function $f(x) = f(x_1, \ldots, x_n)$ on $G = G_1 \times \ldots \times G_n$, where $|G| = g_1 \cdots g_n$ is defined by specifying its values on each $(x_1, \ldots, x_n) \in G$. This can be done in a recursive manner as decision tree as follows. Starting form the root node (level 0) we have g_1 branches for the choices of values for x_1. From each of these g_1 nodes (on level 1), we have g_2 branches for the g_2 choices of the values for x_2. From each node at level $n-1$, we have g_n branches to the leaves of the decision tree. Each leave gives the value of the function corresponding to the element $(x_1, \ldots, x_n)$ of G leading to the leave. Thus derived decision tree is called Multi-terminal decision tree (MTDT), since each node has more than two outgoing branches usually called edges.

A Multi-terminal decision diagram (MTDD) for a given f is derived from the MTDT by sharing isomorphic subtrees and deleting the redundant information in the decision tree [37]. Formally, a MTDD can be defined as follows.

Definition 3.1 *(Multi-terminal Decision Diagram)*
A MTDD for representation of $f \in P(G)$ is a rooted directed acyclic graph $D(V, E)$ with the node set V consisting of non-terminal nodes and terminal or constant nodes. A non-terminal node is labeled with a variable x_i of f and has g_i successors denoted by $succ_k(v) \in V$ with $k \in G_i$. A constant node v is labeled with an element from P and has no successors.

In a MTDT, the i-th level consists of all the nodes to which the variable x_i is assigned. In a MTDD, edges connecting nodes at non successive levels may appear. Cross points are points where such an edge crosses levels in the MTDD. Through cross points, the impact of the deleted nodes from the MTDT is taken into account. The concept of MTDD is explained and illustrated by the following example.

Example 3.6 *Consider a function f on $G_{24} = C_2 \times C_2 \times S_3$, described in the Example 3.5. If f is given by the truth-vector*

$$\mathbf{f} = [0, 6, 2, 1, 0, 0, 2, 1, 1, 0, 0, 0, 1, 1, 1, 1, 1, 1, 1, 1, 1, 1, 2, 2]^T,$$

then it can be represented by the MTDD shown in Figure 3.13. In this figure, x_i^j denotes that the variable x_i takes the value j.

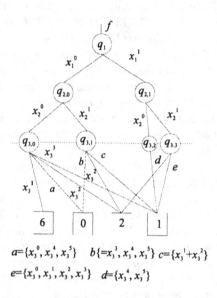

$$a=\{x_3^0, x_3^4, x_3^5\} \quad b\{=x_3^3, x_3^4, x_3^5\} \quad c=\{x_3^1+x_3^2\}$$
$$e=\{x_3^0, x_3^1, x_3^2, x_3^3\} \quad d=\{x_3^4, x_3^5\}$$

Fig. 3.13   MTDD for f in Example 3.6.

3.4.2   FFT on finite non-Abelian groups through DDs

From the theory of Good-Thomas FFT, the calculation of the Fourier transform on a decomposable group G of order g, can be performed through n Fourier transforms on the constituent subgroups G_i of orders g_i.

Calculation of Fourier and Fourier-like transforms through decision diagrams is possible thanks to the recursive structure of decision trees, which is compatible with the recursive structure of Kronecker product representable and related transform matrices.

Calculation procedures based on decision diagrams representations of discrete functions have been proposed for various discrete transforms on Abelian groups [9], [49], [50], [53]. The decision diagrams methods on non-Abelian groups are extensions of these for Abelian groups. Therefore, before discussing decision diagram methods on non-Abelian, we first briefly explain by an example the calculation of spectral transform on Abelian groups.

The following example illustrates the decision diagram methods for calculation of spectral transforms for the case of the Fourier transform on dyadic group of order 8, i.e., the Walsh transform for $n = 3$ whose FFT is shown in Figure 2.3. There are several interpretation of such methods, most of them exploiting the recursive structure of the Walsh matrix [9], [29], [47]. In this example, the method will be explained by pointing relationships to the FFT-like algorithm for the Walsh transform.

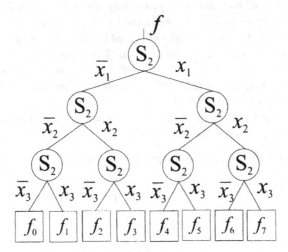

Fig. 3.14 BDD for f in Example 3.7.

Example 3.7 *For $n = 3$, the Walsh transform matrix is defined as*

$$\mathbf{W}(3)^{-1} = \frac{1}{8} \bigotimes_{i=1}^{3} \mathbf{W}(1),$$

where the basic Walsh matrix is $\mathbf{W} = \begin{bmatrix} 1 & 1 \\ 1 & -1 \end{bmatrix}$. For simplicity, in further calculations the scaling factor $\frac{1}{8}$ will be omitted assuming that dividing by 8 will be performed after the spectral coefficients are calculated.

Due to the properties of the Kronecker product, $\mathbf{W}(3)$ can be factorized into the product of three sparse matrices, i.e.,

$$\mathbf{W} = \mathbf{C}_1 \mathbf{C}_2 \mathbf{C}_3,$$

where

$$
\begin{aligned}
\mathbf{C}_1 &= \mathbf{W}(1) \otimes \mathbf{I}_{2\times 2} \otimes \mathbf{I}_{2\times 2}, \\
\mathbf{C}_2 &= \mathbf{I}_{2\times 2} \otimes \mathbf{W}(1) \otimes \mathbf{I}_{2\times 2}, \\
\mathbf{C}_3 &= \mathbf{I}_{2\times 2} \otimes \mathbf{I}_{2\times 2} \otimes \mathbf{W}(1),
\end{aligned}
$$

where $\mathbf{I}_{2\times 2}$ is the (2×2) identity matrix.

The matrices $\mathbf{C}_1$, $\mathbf{C}_2$, and $\mathbf{C}_3$ describe steps in the FFT-like algorithm for the Walsh transform in Figure 2.3, performing the Walsh transform with respect to the variables

x_1, x_2 and x_3. *In a similar way, matrices* $\mathbf{C}_1$, $\mathbf{C}_2$ *and* $\mathbf{C}_3$ *describe processing of nodes in a binary decision tree for* $n = 3$ *in Figure 3.14. The identity matrix means the identical mapping, thus, there is no processing of levels to which they correspond. Thus,* $\mathbf{C}_1$ *shows that we process the root node by* $\mathbf{W}_1$, *and nodes at the other levels remain unprocessed. Similarly, matrices* $\mathbf{C}_2$ *and* $\mathbf{C}_3$ *show that we process by* $\mathbf{W}(1)$ *nodes at levels to which* x_2 *and* x_3 *are assigned. Therefore, it follows that each node in the decision tree should be processed by* $\mathbf{W}(1)$. *The processing means that we perform calculations determined by* $\mathbf{W}(1)$ *over the subfunctions to which point the outgoing edges of the processed node.*

For a function represented by the decision tree in Figure 3.14 the calculations of the Walsh spectrum are performed as follows.

1. *The nodes at the level for* x_3 *are processed as follows:*

$$q_{3,0} = \mathbf{W}(1) \begin{bmatrix} f_0 \\ f_1 \end{bmatrix} = \begin{bmatrix} f_0 + f_1 \\ f_0 - f_1 \end{bmatrix},$$

$$q_{3,1} = \mathbf{W}(1) \begin{bmatrix} f_2 \\ f_3 \end{bmatrix} = \begin{bmatrix} f_2 + f_3 \\ f_2 - f_3 \end{bmatrix},$$

$$q_{3,2} = \mathbf{W}(1) \begin{bmatrix} f_4 \\ f_5 \end{bmatrix} = \begin{bmatrix} f_4 + f_5 \\ f_4 - f_5 \end{bmatrix},$$

$$q_{3,3} = \mathbf{W}(1) \begin{bmatrix} f_6 \\ f_7 \end{bmatrix} = \begin{bmatrix} f_6 + f_7 \\ f_6 - f_7 \end{bmatrix}.$$

2. *The nodes for* x_2 *are processed as*

$$q_{2,0} = \mathbf{W}(1) \begin{bmatrix} q_{3,0} \\ q_{3,1} \end{bmatrix} = \begin{bmatrix} q_{3,0} + q_{3,1} \\ q_{3,0} - q_{3,1} \end{bmatrix}$$

$$= \begin{bmatrix} \begin{bmatrix} f_0 + f_1 \\ f_0 - f_1 \end{bmatrix} + \begin{bmatrix} f_2 + f_3 \\ f_2 - f_3 \end{bmatrix} \\ \begin{bmatrix} f_0 + f_1 \\ f_0 - f_1 \end{bmatrix} - \begin{bmatrix} f_2 + f_3 \\ f_2 - f_3 \end{bmatrix} \end{bmatrix},$$

$$q_{2,1} = \mathbf{W}(1) \begin{bmatrix} q_{3,2} \\ q_{3,3} \end{bmatrix} = \begin{bmatrix} q_{3,2} + q_{3,3} \\ q_{3,2} - q_{3,3} \end{bmatrix}$$

$$= \begin{bmatrix} \begin{bmatrix} f_4 + f_5 \\ f_4 - f_5 \end{bmatrix} + \begin{bmatrix} f_6 + f_7 \\ f_6 - f_7 \end{bmatrix} \\ \begin{bmatrix} f_4 + f_5 \\ f_4 - f_5 \end{bmatrix} - \begin{bmatrix} f_6 + f_7 \\ f_6 - f_7 \end{bmatrix} \end{bmatrix}.$$

3. *The root node for* x_1 *represents the Walsh spectrum since the calculations at this node are as follows:*

$$q_1 = \mathbf{W}(1) \begin{bmatrix} q_{2,0} \\ q_{2,1} \end{bmatrix} = \begin{bmatrix} q_{2,0} + q_{2,1} \\ q_{2,0} - q_{2,1} \end{bmatrix}$$

$$= \begin{bmatrix} \begin{bmatrix} \begin{bmatrix} f_0 + f_1 \\ f_0 - f_1 \end{bmatrix} + \begin{bmatrix} f_2 + f_3 \\ f_2 - f_3 \end{bmatrix} \end{bmatrix} & \begin{bmatrix} \begin{bmatrix} f_4 + f_5 \\ f_4 - f_5 \end{bmatrix} + \begin{bmatrix} f_6 + f_7 \\ f_6 - f_7 \end{bmatrix} \end{bmatrix} \\ \begin{bmatrix} \begin{bmatrix} f_0 + f_1 \\ f_0 - f_1 \end{bmatrix} - \begin{bmatrix} f_2 + f_3 \\ f_2 - f_3 \end{bmatrix} \end{bmatrix} + \begin{bmatrix} \begin{bmatrix} f_4 + f_5 \\ f_4 - f_5 \end{bmatrix} - \begin{bmatrix} f_6 + f_7 \\ f_6 - f_7 \end{bmatrix} \end{bmatrix} \\ \begin{bmatrix} \begin{bmatrix} f_0 + f_1 \\ f_0 - f_1 \end{bmatrix} + \begin{bmatrix} f_2 + f_3 \\ f_2 - f_3 \end{bmatrix} \end{bmatrix} & \begin{bmatrix} \begin{bmatrix} f_4 + f_5 \\ f_4 - f_5 \end{bmatrix} + \begin{bmatrix} f_6 + f_7 \\ f_6 - f_7 \end{bmatrix} \end{bmatrix} \\ \begin{bmatrix} \begin{bmatrix} f_0 + f_1 \\ f_0 - f_1 \end{bmatrix} - \begin{bmatrix} f_2 + f_3 \\ f_2 - f_3 \end{bmatrix} \end{bmatrix} - \begin{bmatrix} \begin{bmatrix} f_4 + f_5 \\ f_4 - f_5 \end{bmatrix} - \begin{bmatrix} f_6 + f_7 \\ f_6 - f_7 \end{bmatrix} \end{bmatrix} \end{bmatrix}$$

$$= \begin{bmatrix} f_0 + f_1 + f_2 + f_3 + f_4 + f_5 + f_6 + f_7 \\ f_0 - f_1 + f_2 - f_3 + f_4 - f_5 + f_6 - f_7 \\ f_0 + f_1 - f_2 - f_3 + f_4 + f_5 - f_6 - f_7 \\ f_0 - f_1 - f_2 + f_3 + f_4 - f_5 - f_6 + f_7 \\ f_0 + f_1 + f_2 + f_3 - f_4 - f_5 - f_6 - f_7 \\ f_0 - f_1 + f_2 - f_3 - f_4 + f_5 - f_6 + f_7 \\ f_0 + f_1 - f_2 - f_3 - f_4 - f_5 + f_6 + f_7 \\ f_0 - f_1 - f_2 + f_3 - f_4 + f_5 + f_6 - f_7 \end{bmatrix}$$

Thus, this is by definition of the Walsh transform, the Walsh spectrum in the Hadamard order. After the calculations are performed, we divide the thus calculated spectral coefficients by 8 to determine their proper values. If each step of the calculation is represented by a decision tree, the result is the decision tree for the Walsh spectrum. Figure 3.15 shows this calculation procedure.

This example illustrates calculations of the Walsh spectrum over the decision tree, since decision trees has the same structure for all the functions for a given number of variables. However, for a given function f, the same calculations can be performed over the decision diagram for f. The reduction of a decision tree into a decision diagram results in the appearance of cross points, which are processed as nodes whose both outgoing edges point to the same subfunction of f.

That approach can be extended to finite non-Abelian groups thanks to the matrix interpretation of the Fourier transform. Calculation through MTDDs consists in the processing of non-terminal nodes and cross points.

A MTDD represents the vector $\mathbf{F}$ for f. Each non-terminal node, or the cross point, in the MTDD can be considered as the root node of a subtree of the MTDD. Each subtree represents a subvector in $\mathbf{F}$. In that way, the processing of nodes and cross points in the MTBDD means calculations over subvectors represented by the subtrees rooted at the nodes where arrive the outgoing edges of the processed nodes or the cross points.

The calculations are performed through some rules that may be conveniently described by matrices. If this is an identity matrix, the processing reduces to the concatenation of subvectors.

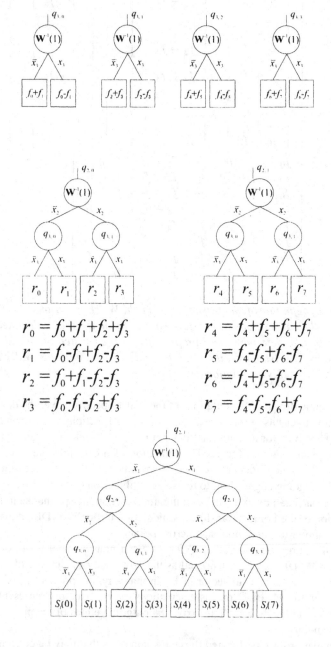

$$r_0 = f_0 + f_1 + f_2 + f_3$$
$$r_1 = f_0 - f_1 + f_2 - f_3$$
$$r_2 = f_0 + f_1 - f_2 - f_3$$
$$r_3 = f_0 - f_1 - f_2 + f_3$$

$$r_4 = f_4 + f_5 + f_6 + f_7$$
$$r_5 = f_4 - f_5 + f_6 - f_7$$
$$r_6 = f_4 + f_5 - f_6 - f_7$$
$$r_7 = f_4 - f_5 - f_6 + f_7$$

Fig. 3.15 MTBDD for the Walsh spectrum for f in Example 3.7.

$$begin \ \{procedure\}$$
$$\text{for } i = n \text{ to } 1$$
$$\text{for } k = 0 \text{ to } Q_i \text{ do}$$

Determine $q_{i,k}$ by using the rule (3.4).

Store $[\mathbf{S}_f] = q_1$.

$$end\{procedure\}$$

Fig. 3.16 Calculation procedure for the Fourier transform.

Definition 3.2 *The operation of concatenation, denoted by $\diamond$, over an ordered set of n vectors $\{\mathbf{A}_1, \ldots, \mathbf{A}_n\}$ of order m is the operation producing a vector $\mathbf{S}$ of order nm consisting of n successive subvectors $\mathbf{A}_1, \ldots, \mathbf{A}_n$.*

Example 3.8 *Application of the operation of concatenation over the set of three vectors $\mathbf{A} = [a_1 a_2 a_3]^T$, $\mathbf{B} = [b_1 b_2 b_3]^T$, $\mathbf{C} = [c_1 c_2 c_3]^T$ produces the vector $\mathbf{D} = \mathbf{A} \diamond \mathbf{B} \diamond \mathbf{C} = [a_1 a_2 a_3 b_1 b_2 b_3 c_1 c_2 c_3]^T$.*

It follows that the calculation of the Fourier coefficients of f on a finite decomposable group G of order g given by the decision diagram can be carried out through the following procedure.

3.4.2.1 *Calculation procedure* Given 3 function f on the decomposable group G of the form (2.4).

1. Represent f by a MTDD. Denote by Q_i the number of non-terminal nodes at the i-th level, i.e., the level corresponding to the variable x_i of f, in the MTDD.

2. Descent the MTDD in a recursive way level by level starting from the constant nodes at $(n + 1)$-st level up to the root node at the level 1.

3. For $i = n$ to 1, process the nodes and cross points by using the rule

$$q_{i,k}(w_i) = r_{w_i} g_i^{-1} \sum_{j=0}^{g_i-1} q_{i+1,j} \mathbf{R}_{w_i}(x_j^{-1}), \tag{3.4}$$

$$k = 0, \ldots, Q_i - 1, w_i = 0, \ldots, K_i - 1,$$

easily derived from the matrix factorization of $[\mathbf{R}]^{-1}$.

Figure 3.16 shows this procedure expressed in a programming pseudo code.

It should be pointed out that there is no matrix computation, and only vector operations are used in the computation of Fourier coefficients through this procedure. This ensures efficiency of the procedure. The vectors are represented by decision diagrams, which permits processing of functions on groups of large orders. The matrix-valued vector determined in the root node q_1 is the Fourier spectrum of f.

The procedure is probably best explained through an example by using the matrix notation.

Example 3.9 *Consider the function f in Example 3.6. The Fourier spectrum for f is calculated as*

$$[\mathbf{S}_f] = [\mathbf{R}_{24}]^{-1}(3)\mathbf{F} \bmod (11).$$

Thus,

$$[\mathbf{S}_f] = 6[5, 9, \mathbf{S}_f(2), 3, 5, \mathbf{S}_f(5), 10, 2, \mathbf{S}_f(8), 7, 1, \mathbf{S}_f(11)]^T \bmod (11),$$

where

$$\mathbf{S}_f(2) = \begin{bmatrix} 5 & 2 \\ 9 & 5 \end{bmatrix}, \mathbf{S}_f(5) = \begin{bmatrix} 5 & 2 \\ 9 & 8 \end{bmatrix}, \mathbf{S}_f(8) = \begin{bmatrix} 9 & 2 \\ 9 & 1 \end{bmatrix}, \mathbf{S}_f(11) = \begin{bmatrix} 1 & 2 \\ 9 & 1 \end{bmatrix},$$

since the scaling factor $6 \cdot 6 \cdot 2$ modulo 11 reduces to 6. Finally,

$$[\mathbf{S}_f] = \begin{bmatrix} 8, 10, \begin{bmatrix} 8 & 1 \\ 10 & 8 \end{bmatrix}, 7, 8, \begin{bmatrix} 8 & 1 \\ 10 & 4 \end{bmatrix}, 5, 1, \begin{bmatrix} 10 & 1 \\ 10 & 6 \end{bmatrix}, 9, 6, \begin{bmatrix} 6 & 1 \\ 10 & 6 \end{bmatrix} \end{bmatrix}^T$$

In (3.1), each matrix $[\mathbf{C}^i]$ describes uniquely one step of the fast Fourier transform performed in n steps. In decision diagrams, the operation in the j-th step of FFT is performed over the nodes and cross points at the j-th level in the decision diagrams. Therefore, the Fourier spectrum of f in Example 3.6 is calculated through MTDD in Figure 3.13 as follows. Note that in this example, all the calculations are in $GF(11)$.

1. *The non-terminal nodes $q_{3,0}, q_{3,1}, q_{3,3}$ and the cross point $q_{3,2}$ are processed first by using the matrix $[\mathbf{S}_3]^{-1}$ in $[\mathbf{R}]^{-1}$. The input data for the procedure are the values of constant nodes. In that way, it is determined*

$$q_{3,0} = [\mathbf{S}_3]^{-1} \begin{bmatrix} 0 \\ 6 \\ 2 \\ 1 \\ 0 \\ 0 \end{bmatrix} = \begin{bmatrix} 7 \\ 3 \\ \begin{bmatrix} 10 & 4 \\ 7 & 2 \end{bmatrix} \end{bmatrix},$$

$$q_{3,1} = [\mathbf{S}_3]^{-1} \begin{bmatrix} 2 \\ 1 \\ 1 \\ 0 \\ 0 \\ 0 \end{bmatrix} = \begin{bmatrix} 8 \\ 8 \\ \begin{bmatrix} 4 & 0 \\ 0 & 4 \end{bmatrix} \end{bmatrix},$$

$$q_{3,2} = [\mathbf{S}_3]^{-1} \begin{bmatrix} 1 \\ 1 \\ 1 \\ 1 \\ 1 \\ 1 \end{bmatrix} = \begin{bmatrix} 1 \\ 0 \\ \begin{bmatrix} 0 & 0 \\ 0 & 0 \end{bmatrix} \end{bmatrix},$$

$$q_{3,3} = [\mathbf{S}_3]^{-1} \begin{bmatrix} 1 \\ 1 \\ 1 \\ 1 \\ 2 \\ 2 \end{bmatrix} = \begin{bmatrix} 5 \\ 7 \\ \begin{bmatrix} 7 & 0 \\ 0 & 4 \end{bmatrix} \end{bmatrix}.$$

2. *The non-terminal nodes* $q_{2,0}, q_{2,1}$ *are processed by using* $\mathbf{W}(1)$. *It is determined*

$$q_{2,0} = \mathbf{W}^{-1}(1) \begin{bmatrix} q_{3,0} \\ q_{3,1} \end{bmatrix} = 6 \left(\begin{bmatrix} 7 \\ 3 \\ \begin{bmatrix} 10 & 4 \\ 7 & 2 \end{bmatrix} \end{bmatrix} + \begin{bmatrix} 8 \\ 8 \\ \begin{bmatrix} 4 & 0 \\ 0 & 4 \end{bmatrix} \end{bmatrix} \right)$$

$$\diamond 6 \left(\begin{bmatrix} 7 \\ 3 \\ \begin{bmatrix} 10 & 4 \\ 7 & 2 \end{bmatrix} \end{bmatrix} + 10 \begin{bmatrix} 8 \\ 8 \\ \begin{bmatrix} 4 & 0 \\ 0 & 4 \end{bmatrix} \end{bmatrix} \right),$$

$$q_{2,1} = \mathbf{W}^{-1}(1) \begin{bmatrix} q_{3,2} \\ q_{3,3} \end{bmatrix} = 6 \left(\begin{bmatrix} 1 \\ 0 \\ \begin{bmatrix} 0 & 0 \\ 0 & 0 \end{bmatrix} \end{bmatrix} + \begin{bmatrix} 5 \\ 7 \\ \begin{bmatrix} 7 & 0 \\ 0 & 4 \end{bmatrix} \end{bmatrix} \right)$$

$$\diamond 6 \left(\begin{bmatrix} 1 \\ 0 \\ \begin{bmatrix} 0 & 0 \\ 0 & 0 \end{bmatrix} \end{bmatrix} + 10 \begin{bmatrix} 5 \\ 7 \\ \begin{bmatrix} 7 & 0 \\ 0 & 4 \end{bmatrix} \end{bmatrix} \right)$$

3. *The root node is processed by using* $\mathbf{W}(1)$. *It is determined*

$$q_1 = \mathbf{W}^{-1}(1) \begin{bmatrix} q_{2,0} \\ q_{2,1} \end{bmatrix} = 6 \left(\begin{bmatrix} 2 \\ 0 \\ \begin{bmatrix} 7 & 2 \\ 9 & 3 \end{bmatrix} \\ 5 \\ 3 \\ \begin{bmatrix} 3 & 2 \\ 9 & 10 \end{bmatrix} \end{bmatrix} + \begin{bmatrix} 3 \\ 9 \\ \begin{bmatrix} 9 & 0 \\ 0 & 2 \end{bmatrix} \\ 9 \\ 2 \\ \begin{bmatrix} 2 & 0 \\ 0 & 9 \end{bmatrix} \end{bmatrix} \right)$$

$$\diamond 6 \left(\begin{bmatrix} 2 \\ 0 \\ \begin{bmatrix} 7 & 2 \\ 9 & 3 \end{bmatrix} \\ 5 \\ 3 \\ \begin{bmatrix} 3 & 2 \\ 9 & 10 \end{bmatrix} \end{bmatrix} + 10 \begin{bmatrix} 3 \\ 9 \\ \begin{bmatrix} 9 & 0 \\ 0 & 2 \end{bmatrix} \\ 9 \\ 2 \\ \begin{bmatrix} 2 & 0 \\ 0 & 9 \end{bmatrix} \end{bmatrix} \right).$$

Thus,

$$q_1 = \left[8, 10, \begin{bmatrix} 8 & 1 \\ 10 & 8 \end{bmatrix}, 7, 8, \begin{bmatrix} 8 & 1 \\ 10 & 4 \end{bmatrix}, 5, 1, \begin{bmatrix} 10 & 1 \\ 10 & 6 \end{bmatrix}, 9, 6, \begin{bmatrix} 6 & 1 \\ 10 & 6 \end{bmatrix} \right]^T$$
$$= [\mathbf{S}_f].$$

Thus, the Fourier spectrum of f is equal to the matrix-valued vector determined in q_1 and it is equal to that calculated by definition of the Fourier transform.

Each step of the calculation can be represented by a MTDD, which results in the MTDD for the Fourier spectrum of f. Figure 3.17 shows the MTDD for the Fourier spectrum generated by using the proposed procedure.

3.4.3 MTDDs for the Fourier spectrum

A MTDD for the Fourier spectrum differs from that representing f in the same way as the FFT algorithms on Abelian groups differ from FFT on non-Abelian groups [41]. Decision diagrams representing the Fourier spectrum of f on finite non-Abelian groups are matrix-valued, since the values of constant nodes are the Fourier coefficients. The number of outgoing edges of nodes at the i-th level is determined by the cardinality K_i of the dual object Γ_i of G_i. In the MTDD for f, the number of outgoing edges of nodes at the i-th level is equal to the order g_i of G_i.

Efficiency of the MTDD representation of f depends on the number of different values f can take. In the same way, the efficiency of DDs representation of the Fourier spectrum of f depends on the number of different Fourier coefficients. In this way, it depends indirectly on the number of different values of f.

In a matrix-valued MTDD (mvMTDD) for $\mathbf{S}_f$, the matrices representing values of constant nodes can be represented by MTDDs [43], in the same way as any matrix can be represented by a MTDD [9]. In that way, the number-valued MTDDs (nvMTDDs) are derived [43].

Figure 3.18 shows a nvMTBDD for $\mathbf{S}_f$ in Example 3.9. It is derived from the mvMTDD in Figure 3.17. The columns of matrices representing values of constant nodes are written as subvectors in a vector, which is then represented by a MTDD. In this figure, to make it clear, some constant nodes are repeated.

Thanks to the spectral interpretation of DDs [48], MTDDs for the Fourier spectrum of f can be interpreted as Fourier DDs for f [42], [43], [44].

3.4.4 Complexity of DDs calculation methods

The structure of a FFT is described by the number of levels and by connections of nodes within levels. For a given G, the structure of the FFT is determined by the assumed decomposition into the product of subgroups G_i of smaller orders (2.4). For the assumed decomposition, the FFT on G has the same structure, thus the same complexity, for any f. Therefore, in calculation of the Fourier transform through the FFT we do not take into account any peculiar properties a function may have.

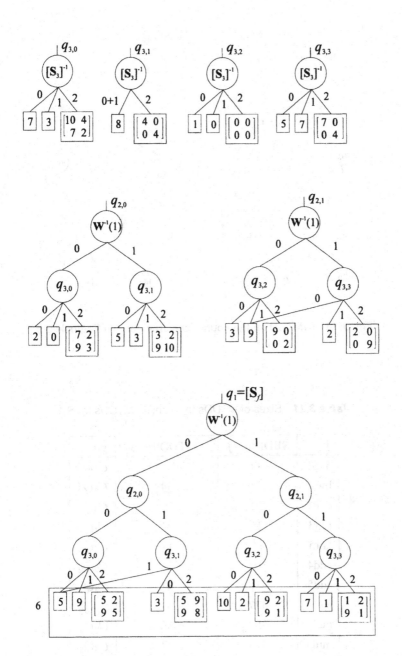

Fig. 3.17 Calculation of the Fourier spectrum through MTDD.

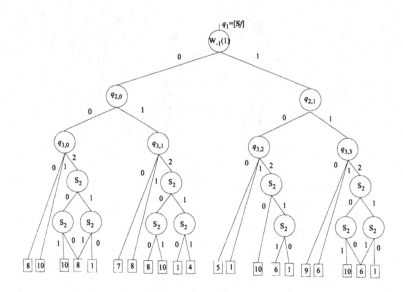

Fig. 3.18 nvMTDD for the Fourier spectrum for f in Example 3.9.

Table 3.11 Sizes of MTDDs for f and nvMTDDs for S_f.

f	SBDD for f	nvMTDD for S_f	group
5xp1	90	167	$C_2 Q_2^2$
bw	116	34	$C_4 Q_2^2$
con1	20	25	$C_2 Q_2^2$
rd53	25	31	$C_4 Q_2$
xor5	11	11	$C_4 Q_2^2$
add4	103	28	$C_2 Q_2$
add5	226	34	$C_4 Q_2$
add6	477	33	Q_2^2
mul4	159	66	$C_2 Q_2$
mul5	473	87	$C_4 Q_2$
mul6	788	94	Q_2^2

Compactness of a decision diagram for f is based upon deleting isomorphic parts in the decision tree for f. Thus, in calculation of the Fourier transform through DD for f, we do not repeat calculations over identical subvectors in the vector representing values of f. In that way, unlike FFT, we take into account properties of the processed functions.

At each node and the cross point at the i-th level, we perform calculations determined by the Fourier transform on G_i. Thus complexity of calculations through DDs is proportional to the number of nodes and cross points in the DD for f, usually denoted as the size of the DD. If a function f has some peculiar properties, as for example symmetry, or decomposability, then the MTDD for f has smaller size, and calculation of the Fourier transform is simpler. In reporting experimental results, the size of a DD is usually considered as the number of non-terminal nodes. It is assumed that for an arbitrary function, the number of cross points is smaller than $30 - 40\%$ of the number of non-terminal nodes [37].

The same as in the FFT, the complexity of calculations depends on the complexity of calculation rules derived from the Fourier transforms on G_i.

Table 3.11 shows the sizes of MTDDs for some *mcnc* benchmark switching functions and their Fourier transforms. In this table, the multi-output switching functions are represented by Shared binary DDs (SBDDs) [37]. The domain group is of the form C_2^n. For non-Abelian groups, the assumed decompositions are shown. C_2 and C_4 are the cyclic groups of orders 2 and 4, and Q_2 denotes the quaternion group of order 8. The price for the reduced size is the increased number of outgoing edges of nodes. The advantage is the reduced depth of the decision diagrams.

REFERENCES

1. Agaian, S., Astola, J., Egiazarian, K., *Binary Polynomial Transforms and Nonlinear Digital Filters*, Marcel Dekker, New York, 1995.

2. Apple, G., Wintz, P., "Calculation of Fourier transforms on finite Abelian groups", *IEEE Trans.*, Vol. IT-16, 1970, 233-234.

3. Beauchamp, K.G., *Applications of Walsh Series and Related Functions: With an Introduction to Sequency Theory*, Academic Press, New York, 1984.

4. Besslich, Ph.W., "Efficient computer method for XOR logic design", *IEE Proc., Part E*, Vol. 129, 1982, 15-20.

5. Besslich, Ph.W., "Spectral processing of switching functions using signal flow transformations", in Karpovsky, M.G., (Ed.), *Spectral Techniques and Fault Detection*, Academic Press, Orlando, FL, 1985.

6. Bryant, R.E. "Graph-based algorithms for Boolean functions manipulation", *IEEE Trans. Comput.*, Vol. C-35, No. 8, 1986, 667-691.

7. Burrus, C.S., Parks, T.W., *DFT/FFT and Convolution Algorithms: Theory and Implementation*, John Wiley, New York, 1985.

8. Cairns, T.W., "On the fast Fourier transform on finite Abelian groups", *IEEE Trans. on Computers*, Vol. C-20, 1971, 569-571.

9. Clarke, E.M., McMillan, K.L., Zhao, X., Fujita, M., "Spectral transforms for extremely large Boolean functions", in Kebschull, U., Schubert, E., Rosentiel, W., Eds., *Proc. IFIP WG 10.5 Workshop on Applications of the Reed-Muller Expansion in Circuit Design*, 16-17.9.1993, Hamburg, Germany, 86-90.

10. Cooley, J.W., Tukey, J.W., "An algorithm for the machine calculation of complex Fourier series", *Math. Computation*, Vol. 19, 1965, 297-301.

11. Depeyrot, M., "Fondements algebriques de la transformation de Fourier rapide", 1E/11, Centre d'Automatique, Ecole des Mines, Fontainebleau, France, 1969.

12. Egner, S., Püschel, M., "Generation of fast discrete signal transforms", *IEEE Trans. on Signal Processing*, Vol. 49, No. 9, 2001, 1992-2002.

13. Frigo, M., Johnson, S.G., "The fastest Fourier transform in the West", MIT-LCS-R-728, Massachusetts Institute of Technology, Boston, MA, September 11, 1997.

14. Gibbs, J.E., Stanković, R.S., "Matrix interpretation of Gibbs derivatives on finite groups", private communication, 1988.

15. Good, I.J., "The interaction algorithm and practical Fourier analysis", *J. Roy. Statist. Soc.*, ser. B, Vol.20, 361-372, Addendum, Vol.22, 1960, 372-375.

16. Good, I.J., "The relationship between two fast Fourier transforms", *IEEE Trans. Comput.*, Vol. C-20, No. 3, 310-317, 1971.

17. Harmuth, H.F., *Sequency Theory: Foundations and Applications*, Academic Press, New York, 1977.

18. Heideman, M.T., Johnson, D.H., Burrus, C.S., "Gauss and the history of the fast Fourier transform", *Archiv for History of Exact Science*, Vol. 34, No. 3, 1985, 265-277. Also in *IEEE ASSP Magazine*, Vol. 1, No. 4, Oct. 1984, 14-21.

19. Hurst, S.L., Miller, D.M., Muzio, J.C., *Spectral Techniques in Digital Logic*, Academic Press, Toronto, 1985.

20. Hurst, S.L., *The Logical Processing of Digital Signals*, Crane Russak and Edvard Arnold, Basel and Bristol, 1978.

21. Johanson, J., Püschel, M., "In search of the optimal Walsh-Hadamard transform", *Proc. Int. Conf. on Acoustics, Speech and Signal Processing (ICASSP)*, 2000, 3347-3350.

22. Karpovsky, M.G., *Finite Orthogonal Series in the Design of Digital Devices*, Wiley, New York, 1976.

23. Karpovsky, M.G., "Fast Fourier transforms on finite non-Abelian groups", *IEEE Trans. Comput.*, Vol. C-26, No. 10, Oct. 1977, 1028-1030.

24. Karpovsky, M.G., Trachtenberg, E.A., "Some optimization problems for convolution systems over finite groups", *Inform. Control.*, 34, 1977, 227-247.

25. Karpovsky, M.G., Trachtenberg, E.A., "Fourier transform over finite groups for error detection and error correction in computation channels", *Inform. Control*, Vol. 40, No. 3, 1979, 335-358.

26. Karpovsky, M.G., Trachtenberg, E.A., "Statistical and computational performance of a class of generalized Wiener filters", *IEEE Trans. Information Theory*, Vol. IT-32, May 1986, 303-307.

27. Kumhom, P., *Design, Optimization and Implementation of a Universal FFT Processor*, Ph.D. thesis, Drexel University, Philadelphia, PA, 2001.

28. Kumhom, P., Johnson, J.R., Nagvajara, P., "Design, optimization, and implementation of a universal FFT processor", *Proc. 13th IEEE Int. ASIC/SOC Conference*, 2000, 182-186.

29. Miller, D.M., "Spectral transformation of multiple-valued decision diagrams", *Proc. 24th Int. Symp. on Multiple-Valued Logic*, Boston, MA, 22.-25.5.1994, 89-96.

30. Moraga, C., "On some applications of the Chrestenson functions in logic design and data processing", *Mathematics and Computers in Simulation*, 27, 1985, 431-439.

31. Nussbaumer, H.J., *Fast Fourier Transform and Convolution Algorithms*, Springer-Verlag, Berlin, Heidelberg, New York, 1981.

32. Park, N., Prasanna, V., "Cache conscious Walsh-Hadamard transform", *Proc. Int. Conference on Acoustics, Speech, and Signal Processing (ICASSP 2001)*, Vol. 2, 2001.

33. Roziner, T.D., Karpovsky, M.G., Trachtenberg, L.A., "Fast Fourier transforms over finite groups by multiprocessor system", *IEEE Trans.*, Vol. ASSP-38, No. 2, 1990, 226-240.

34. Sasao, T., "Representations of logic functions by using EXOR operators", *Proc. IFIP WG 10.5 Workshop on Application of the Reed-Muller Expansion in Circuit Design*, 27-29.8.1995, Makuhari, Chiba, Japan, 11-20.

35. Sasao, T., "Representations of logic functions by using EXOR operators", in Sasao, T., Fujita, M., (eds.), *Representations of Discrete Functions*, Kluwer Academic Publishers, Boston, 1996, 29-54.

36. Sasao, T., *Switching Theory for Logic Synthesis*, Kluwer Academic Publishers, Boston, 1999.

37. Sasao, T., Fujita, M., *Representations of Discrete Functions*, Kluwer Academic Publishers, Boston, 1996.

38. Singer, B.W., *Automating the Modeling and Optimization of the Performance of Signal Processing Algorithms*, Ph.D. thesis, CMU-CS-01-156, School of Computer Science, Computer Science Department, Carnegie Melon University, Pittsburgh, PA, December 2001.

39. Srinivasan, A., Kam, T., Malik, Sh., Brayant, R.K., "Algorithms for discrete function manipulation", in *Proc. Inf. Conf. on CAD*, 1990, 92-95.

40. Stanković, R.S., "Matrix interpretation of fast Fourier transform on finite non-Abelian groups", *Res. Rept. in Appl. Math.*, YU ISSN 0353-6491, Ser. Fourier Analysis, Rept. No. 3, April 1990, 1-31, ISBN 86-81611-03-8.

41. Stanković, R.S., "Matrix interpretation of the fast Fourier transforms on finite non-Abelian groups", *Proc. Int. Conf. on Signal Processing, Beijing/90*, 22.-26.10.1990, Beijing, China, 1187-1190.

42. Stanković, R.S., "Fourier decision diagrams for optimization of decision diagrams representations of discrete functions", *Proc. Workshop on Post Binary-Ultra Large Scale Integration*, Santiago de Campostela, Spain, 1996, 8-12.

43. Stanković, R.S., "Fourier decision diagrams on finite non-Abelian groups with preprocessing," *Proc. 27th Int. Symp. on Multiple-Valued Logic*, Antigonish, Nova Scotia, Canada, May 1997, 281-286.

44. Stanković, R.S., *Spectral Transform Decision Diagrams in Simple Questions and Simple Answers*, Nauka, Belgrade, 1998.

45. Stanković, R.S., Milenović, D., "Some remarks on calculation complexity of Fourier transforms on finite groups", *Proc. 14th European Meeting on Cybernetics and Systems Research, CSMR'98*, April 15-17, 1998, Vienna, Austria, 59-64.

46. Stanković, R.S., Milenović, D., Janković, D., "Quaternion groups versus dyadic groups in representations and processing of switching fucntions", *Proc. 20th Int. Symp. on Multiple-Valued Logic*, Freiburg im Breisgau, Germany, May 20-22, 1999, 19-23.

47. Stanković, R.S., Stanković, M., Janković, D., *Spectral Transforms in Switching Theory, Definitions and Calculations*, Nauka, Belgrade, 1998.

48. Stanković, R.S., Sasao, T., Moraga, C., "Spectral transform decision diagrams", *Proc. IFIP WG 10.5 Workshop on Application of the Reed-Muller Expansion in Circuit Design*, 27-29.8.1995, Makuhari, Chiba, Japan, 46-53.

49. Stanković, R.S., Stanković, M., Moraga, C., Sasao, T., "Calculation of Vilenkin-Chrestenson transform coefficients of multiple-valued functions through multiple-place decision diagrams", *Proc. 5th Int. Workshop on Spectral Techniques*, March 15-17, 1994, Beijing, China, 107-116.

50. Stanković, R.S., Stanković, M., Moraga, C., Sasao, T., "Calculation of Reed-Muller-Fourier coefficients of multiple-valued functions through multiple-place decision diagrams", *Proc. 24th Int. Symp. on Multiple-valued Logic*, Boston, MA, May 22-25, 1994, 82-88.

51. Stanković, R.S., Stojić, M.R., Stanković, M.S., (eds.), *Recent Developments in Abstract Harmonic Analysis with Applications in Signal Processing*, Nauka, Belgrade and Elektronski fakultet, Niš, 1996.

52. Thomas, L.H., "Using a computer to solve problems in physics", in *Application of Digital Computers*, Ginn, MA, 1963.

53. Thornton, M.A., "Modified Haar transform calculation using digital circuit output probabilities", *Proc. IEEE Int. Conf. on Information, Communications and Signal Processing* (1st ICICS), Singapore, Vol. 1, September 1997, 52-58.

54. Trachtenberg, E.A., "Applications of Fourier Analysis on Groups in Engineering Practices", in Stanković, R.S., Stojić, M.R., Stanković, M.S., (eds.), *Recent Developments in Abstract Harmonic Analysis with Applications in Signal Processing*, Nauka, Belgrade and Elektronski fakultet, Niš, 1996, 331-403.

55. Trachtenberg, E.A., Karpovsky, M.G., "Filtering in a communication channel by Fourier transforms over finite groups", in M.G. Karpovsky, (ed.), *Spectral Techniques and Fault Detection*, , Academic Press, Orlando, FL, 1985.

56. Varma, D., Trachtenberg, L.A., "Efficient spectral techniques for logic synthesis", in T. Sasao, Ed., *Logic Synthesis and Optimization*, Kluwer Academic Publishers, Boston, 1993, 215-232.

4

Optimization of Decision Diagrams

In Section 3.4, MTDTs are used as a data structure which permits efficient calculation of the Fourier transform of functions defined on groups of large orders. In this Chapter, we discuss application of non-Abelian groups in the optimization of decision diagram representations for discrete functions, including switching and multiple-valued (MV) functions as particular examples. The presentation is mainly oriented towards decision diagrams for switching functions, since they are the most often met in practice. However, generalizations to multiple-valued, integer or complex-valued functions is simple.

In the theory of decision diagram representations [17], [23], decision diagrams are derived by reducing the corresponding decision trees (DTs). The reduction is done by deleting or sharing redundant nodes in a DT depending on the equal or otherwise assignment of the decision variables assigned to the non-terminal nodes in the DT. Reduction is performed through rules suitably formulated depending on the range of represented functions and the nodes used in definition of the decision trees. In that respect, BDD reduction rules, zero-suppressed BDD reduction rules [16], and generalized BDD reduction rules [27] are distinguished.

There are two general approaches to the optimization of decision diagram representations of switching functions reported in the literature:

1. Given a switching function f, represent it by a binary decision diagram (BDD) [3]. Do the optimization by changing the meaning of nodes , i.e., by choosing among the Shannon, positive and negative Davio nodes [16]. In that way BDDs are transferred into Kronecker or pseudo-Kronecker decision diagrams. In the spectral interpretation, that means the optimization by searching among different bases to define the decision tree best suited to the given f [27].

2. Given a switching function f, represent it by a BDD. Do the optimization of the representation by coding pairs of binary variables by four-valued variables. In that way, the BDD representation is transferred into a Quaternary decision diagram (QDD) representation [18]. Further optimization can be done by applying the first approach of optimization to quaternary decision diagrams, which produces quaternary Kronecker and pseudo-Kronecker DDs [18].

A disadvantage of the first method is that it does not permit reduction of levels in the decision tree, except in some special cases when the represented functions possess some particular symmetry properties. That was the motivation for the second approach based on recoding of subsets of variables [18].

For reasons of the intended practical realizations, the method was restricted to pairs of variables [18], although the extension to any subset of variables can be directly given. Moreover, the method does not relate to the function values and can be generalized to decision diagrams for discrete functions, thanks to the extension of the concept of BDDs [3] into multi-terminal decision diagrams (MTBDDs) [5].

A drawback of this approach to the optimization of decision diagram representations is that it requires nodes with many outgoing edges. Such nodes change the reduction possibilities in decision trees and usually increase the number of nodes per levels in the derived decision diagrams. The reduction of the number of levels and nodes per level in a decision diagram are thus two opposite requirements.

In what follows, we first discuss the problem and give a group-theoretic interpretation of the optimization of decision diagrams by coding subsets of variables. Then we offer a solution of the problem by proposing the use of Fourier decision diagrams on finite non-Abelian groups.

4.1 REDUCTION POSSIBILITIES IN DECISION DIAGRAMS

Definition 3.1 of MTDDs can be further elaborated as follows.

Definition 4.1 *A decision diagram (DD) that represents an n-variable discrete function* $f(x_1, \ldots, x_n)$ *is a rooted acyclic graph* $G = (V, E)$ *with the edge set* E *and the node set* V *consisting of non-terminal and constant nodes. A variable* x_i *is assigned to each non-terminal node* $v \in V$ *and is called the decision variable for* v. *If* x_i *takes* g_i *different values, then* v *has* g_i *outgoing edges labeled with the* g_i *values. All the nodes assigned to the same variable* x_i *form the i-th level in the DD, assuming the same order of variables along the paths from the root node to the constant nodes.*

Decision diagrams with the same order of variables in all the paths are usually called ordered decision diagrams. Notice that there are also defined Free binary decision diagrams (FBDDs) [2], where the order of variables along different paths may be different. However, the restriction that a variable cannot appear several times in a path is preserved in FBDDs.

For a given function f, a decision diagram is derived by the reduction of the corresponding decision tree (DT) that is generated by a recursive application of some

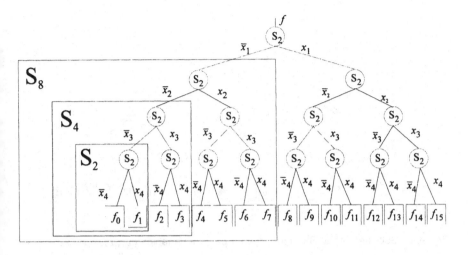

Fig. 4.1 Shannon tree for $n = 4$.

expansion rule with respect to each variable x_i of f. The reduction is done by sharing the isomorphic subtrees in the decision trees and by deleting the redundant nodes in the decision trees. A node is redundant if all its outgoing edges point to the same node at a successing level in the decision tree. Therefore, a decision tree is the basic concept in decision diagram representations of discrete functions. In the following discussion, the structure of a decision tree is an important concept [23].

Observation 4.1 *The structure of a DT is determined by the number of levels and the number of outgoing edges of non-terminal nodes.*

Note that the number of outgoing edges of nodes at the i-th level uniquely determines the number of nodes at the $(i+1)$-st level in a decision tree. Therefore, we may alternatively say that the structure of the decision tree is determined by the number of levels and the number of nodes per levels. These two statements are equivalent.

Binary decision diagrams (BDDs) are derived by the reduction of the Shannon trees (Figure 4.1) by BDD reduction rules [16]. For an n-variable function, the Shannon tree consists of n levels with non-terminal nodes with two outgoing edges. In Figure 4.1, and the following figures, the nodes where the Shannon decomposition or its generalizations are performed, are denoted by S_i, with index i denoting the number of outgoing edges. The labels at the edges and the values of constant nodes are not shown in the figures, if their meaning is obvious form the context.

Quaternary decision diagrams (QDDs) [18] are derived by recoding of pairs of switching variables by four-valued variables, $(x_i, x_j) = X_k$, $x_i, x_j \in \{0, 1\}$, $X_k \in \{0, 1, 2, 3\}$.

In the Shannon tree (schematic representation in Figure 4.2), that means the replacement of each subtree shown in Figure 4.3(a) by a single node with four outgoing edges (Figure 4.3(b)). In that way, the Shannon tree (Figure 4.1) is replaced by the

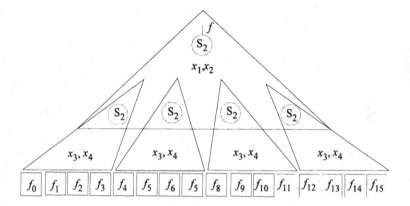

Fig. 4.2 Subtrees in the Shannon tree for pairs of variables $(x_1, x_2), (x_3, x_4)$.

quaternary decision tree (QDT) (Figure 4.4) with the number of non-terminal levels reduced to a half compared to the starting Shannon tree (two non-terminal levels in QDT instead of four non-terminal levels in the Shannon tree in this example). However, in QDDs nodes with four outgoing edges are required.

The method can be extended to the coding of any subset of variables. For example, the Shannon tree for $n = 4$ may be decomposed into subtrees corresponding to x_1 and $(x_2, x_3, x_4) = X_2$, $X_2 \in \{0, \ldots, 7\}$ (Figure 4.5). In this way, the subtree in Figure 4.6(a) is replaced by a single node with eight outgoing edges (Figure 4.6(b)). Thus, the Shannon tree for $n = 4$ is replaced by the tree in Figure 4.7. In this tree, besides the number of levels, the number of non-terminal nodes is also reduced. However, nodes with eight outgoing edges are required. Alternatively, the subtree in Figure 4.6(a) may be replaced by the subtree in Figure 4.8 consisting of a node with two outgoing edges and two nodes with four outgoing edges. The resulting decision tree is shown in Figure 4.9.

A drawback of this method for optimization of Decision diagram representations is the appearance of non-terminal nodes with considerable number of outgoing edges. In that way, by recoding subset of variables, we are changing the structure of the decision tree.

Observation 4.2 *In a decision tree, the number of outgoing edges of nodes at the i-th level determine the number of nodes at the $(i + 1)$-th level and in that way determine the reduction possibilities in the decision tree. By changing the structure of the decision tree, we change the reduction possibilities in the decision tree.*

To explain this statement, denote by $Z_{16} = [z_1, \ldots, z(15)]^T$, the vector representing the values of the constant nodes in the Shannon tree for $n = 4$. A node at the 4-th level, thus, corresponding to x_4, may be deleted if the corresponding subvector of Z_{16}, $Z_0 = [z_0, z_1]^T, \ldots, Z_7 = [z_{14}, z_{15}]^T$ is a constant vector. A node at the 4-th level may be shared if the corresponding subvectors $Z_i, Z_j, i \neq j, i, j \in \{0, \ldots, 7\}$

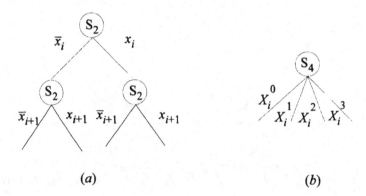

Fig. 4.3 (a) Subtree in the Shannon tree for a pair of variables (x_i, x_{i+1}), (b) QDD non-terminal node.

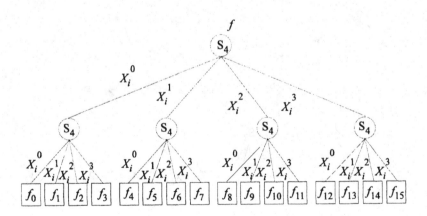

Fig. 4.4 QDD for $n = 4$.

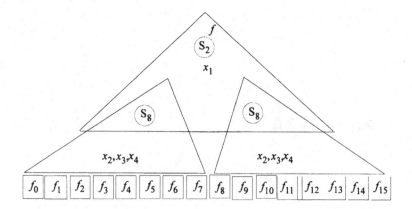

Fig. 4.5 Subtrees in the Shannon tree for $x_1, (x_2, x_3, x_4)$.

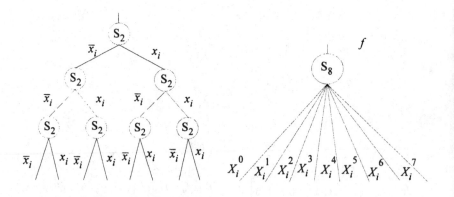

Fig. 4.6 (a) Subtree in the Shannon tree for (x_{i-1}, x_i, x_{i+1}), (b) Non-terminal node with eight outgoing edges.

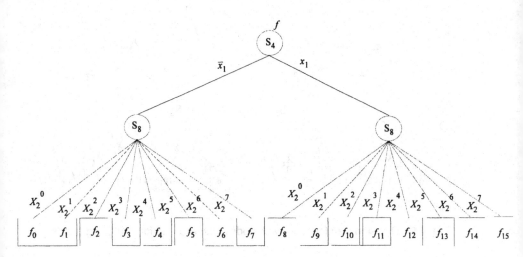

Fig. 4.7 Decision tree with nodes with two and eight outgoing edges.

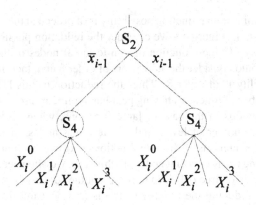

Fig. 4.8 Subtree with the Shannon S_2 and QDD non-terminal nodes.

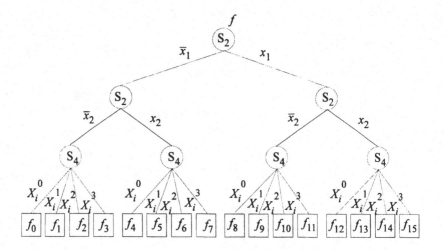

Fig. 4.9 Decision tree for $n = 4$ with nodes with two and four outgoing edges.

are equal. We call this property as the reduction possibility of order 2. It is true generally for the nodes at the n-th level in any decision tree whose structure is equal to the structure of the Shannon tree for any n. Examples are the positive Davio tree and Reed-Muller tree and other related decision trees [16]. To delete or share a node at the level $(n - 1)$ in such DT we should consider the reduction possibility of order 4 defined as above, but for subvectors of order 4. Further we should consider the reduction possibility of order 2^{n-1} for the nodes at the second level in the Shannon tree.

In QDDs, the minimum reduction possibility is of order 4 at the $n/2$-th level, while in the decision tree in Figure 4.9 we consider the reduction possibilities of orders 4 and 8, respectively. In the reduction of non-terminal nodes in the decision tree in Figure 4.7 we should consider the subvectors of order 8 and, therefore, the minimum reduction possibility is of order 8. Thus, the reduction of this DT may be done in only a small number of functions having peculiar symmetry properties. The problem is much more obvious in the case of large functions, when decision trees consist of a considerable number of levels and non-terminal nodes. The problem is also very hard in representations of discrete functions taking a large number of different values, thus, having many constant nodes. In this case, we often need almost complete multi-terminal decision tree (MTDT) to represent the function.

Therefore, by reducing the number of levels in a decision tree by recoding the subsets of variables, i.e., by using the nodes with the increased number of outgoing edges, we decrease the probability to delete or share a non-terminal node in the decision tree.

Looking in the opposite direction, from the top of the decision tree, we are increasing the number of non-terminal nodes at the second level from 2 in the Shannon trees to 4 in QDDs. Since we are increasing the order of the reduction possibility, we

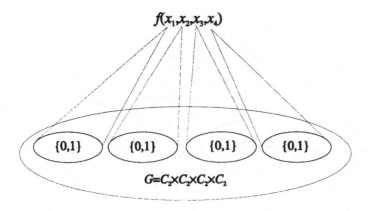

Fig. 4.10 Decomposition of the domain group $G_{16} = C_2^4$.

are increasing the number of non-terminal nodes per levels in the resulting decision diagram. It follows that the reduction of levels and the number of nodes per levels are contradictory requirements in the optimization of decision diagrams.

The same consideration applies to the generalization of decision diagrams for switching functions into decision diagrams for discrete functions. BDDs are generalized into multi-terminal binary decision diagrams (MTBDDs) by allowing integers or complex numbers as the values of constant nodes [5]. Thus, BDD and MTBDDs are derived from decision trees of the same structure, but differ in the values of constant nodes. In the same way QDDs may be generalized to represent integer or complex-valued functions. The same can be done with other decision diagrams defined through coding of different subsets of variables. In that way we have multi-terminal decision diagrams (MTDDs) consisting of nodes with arbitrary number of outgoing edges.

In what follows, we give a group-theoretic interpretation of decision trees for switching and discrete functions to express clearly where the problems in reduction of levels and nodes of a decision diagram originate from and to eventually find a way to solve them.

4.2 GROUP-THEORETIC INTERPRETATION OF DD

In decision diagram representation of an n-variable switching functions f by the Shannon tree, f is considered as a function on the finite dyadic group of order 2^n, $C_2^n = \times_{i=1}^n C_2$, where $C_2 = (\{0,1\}, \oplus)$, and $\oplus$ denotes the modulo 2 addition (EXOR). Thus, for $n = 4$ the domain group G of f is $G = C_2^n = C_2 \times C_2 \times C_2 \times C_2$ (Figure 4.10). In QDD representation (Figure. 4.4), through the coding of pairs of variables we change the decomposition of the domain group G of f into the product of two cyclic groups of order four, i.e., $G = C_4 \times C_4$ (Figure 4.11).

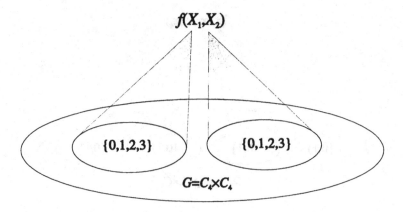

Fig. 4.11 Decomposition of the domain group $G_{16} = C_4^2$.

In the group theoretic approach to discrete functions, the structure of decision trees is determined by the structure of the group G on which a given f is considered. Therefore, the Statement 1 can be reformulated as follows.

Observation 4.3 *The structure of a MTDT is determined by the decomposition of the domain group G of f. The number of levels in the decision tree is equal to the number of subgroups G_i into which the domain group G of f is decomposed as a direct product. The number of outgoing edges of nodes at the i-th level and, thus, the number of nodes at the $(i + 1)$-st level is determined by the order g_i of G_i. The number of constant nodes is equal to the order g of G.*

In a decision tree, by changing the meaning of nodes, we change the order of elements in the vector Z_g representing the values of constant nodes. In that way, we may increase the probability of finding identical subvectors in Z_g and to reduce the number of nodes in the resulting decision diagram. However, we can not reduce the number of levels, since we do not change the structure of the basic decision tree we are starting from.

The same applies to the MTBDDs, Walsh decision diagrams (WDDs), Arithmetic transform decision diagrams (ACDDs) [27] and binary moment decision diagrams (BMDDs) [4], which differ from ACDDs in the reduction rules [27].

As is noted above, the reduction of levels may be achieved by recoding the subset of variables, since in this way we are changing the structure of decision trees we are starting from. From Statement 4.3, in the group-theoretic approach we have the following statement.

Observation 4.4 *The number of levels in a decision diagram may be reduced by using different decompositions of the domain group G of f into fewer but larger subgroups.*

For example, in QDDs we are using the subgroups C_4 of order four (Figure 4.11), instead the subgroups C_2 of order two (Figure. 4.10). Besides the thus achieved reduction of levels, further reduction may be done by changing the meaning of nodes. In that way the quaternary pseudo Kronecker decision diagrams are introduced [18]. The same can be done for decision diagrams corresponding to any other decomposition of G.

However, as noted above, the use of large subgroups requires nodes with considerable number of outgoing edges. Such nodes decrease the probability for reduction.

We recall the spectral interpretation of decision diagrams [27] to look for a solution of this problem caused by the contradictory requirements of reducing both levels and nodes per levels.

In a decision diagram, the meaning of nodes determines the labels at the outgoing edges of the node. For example, the labels of outgoing edges of the Shannon, positive Davio and negative Davio nodes are $\{\overline{x}_i, x_i\}$, $\{1, x_i\}$ and $\{1, \overline{x}_i\}$, respectively [16]. In Walsh transform DDs (WDDs), the labels at the edges are $\{1, 1 - 2x_i\}$ [27]. In QDDs, the labels at the edges are $\{X_i^0, X_i^1, X_i^2, X_i^3\}$, where X_i^j denotes that X_i takes the value j [18].

In spectral interpretation of decision diagrams, we refer to the labels at outgoing edges of nodes. We say that decision trees are the graphical representations of the Fourier and Fourier-like expansions on finite groups with respect to some particular bases. The basis functions are determined as products of labels at the edges in each path from the root node to the constant nodes in the DT. We speak of the Fourier expansions if we use as basis functions the group characters for Abelian and unitary irreducible representations for non-Abelian groups. For any other basis we say the Fourier-like expansion.

In this interpretation, the values of constant nodes in decision trees are the Fourier or Fourier-like spectral coefficients of f.

A Fourier or Fourier-like spectrum can be considered as a function on the dual object Γ of G [10], [11]. The Fourier transform on a finite decomposable group G of order g can be expressed as the composition of n Fourier transforms on the constituent subgroups G_i of orders g_i. Therefore, we have the following observation for the Fourier decision diagrams.

Observation 4.5 *The number of constant nodes in a Fourier DT is equal to the cardinality K of the dual object Γ of G. The number of the outgoing-edges of nodes at the i-th level is equal to the cardinality K_i of the dual object Γ_i of G_i.*

This observation is not in contradiction with the Observation 4.4. MTBDDs and MTDDs are defined with respect to the identical mapping and, thus, in this case the domain group G and its dual object coincides. Moreover, the dual object Γ of any Abelian group is isomorphic to G. Therefore, the Statement 4.4 applies to any decision tree on Abelian groups. Statement 4.5 is a generalization involving various decision trees on both Abelian and non-Abelian groups as the domain groups for f.

The advantages of decomposition of the domain group G into non-Abelian subgroups are in the following. For non-Abelian groups, it is always $K_i < g_i$, since at least one of the unitary irreducible representations R_w has the order $r_w > 1$. Therefore, we can define the Fourier decision trees by using the unitary irreducible representations as the basis. In these decision trees, the number of outgoing edges of nodes at the i-th level is equal to K_i instead of g_i in the case of Fourier and any other decision tree on Abelian groups. Therefore, with Fourier decision diagrams, derived by the reduction of Fourier decision trees, we can use large subgroups G_i in the decomposition of G to reduce the number of levels. If these subgroups are non-Abelian groups, DT will consists of nodes with a smaller number of outgoing edges compared to any DT on Abelian groups. Thus, with Fourier decision diagrams on non-Abelian groups [22], we can reduce both number of levels and nodes per levels in a decision tree.

4.3 FOURIER DECISION DIAGRAMS

4.3.1 Fourier decision trees

As it was shown in Section 2.5, the Fourier transform on a finite group G decomposable in the form (2.4) may be considered as the n-dimensional Fourier transform on the constituent subgroups G_i. Due to that, the fast calculation algorithm are defined, in which each step performs the Fourier transform on a constituent subgroup G_i of the domain group G. In Example 3.7, it is explained the relationship between steps of FFT-like algorithms and decision diagrams and it is shown how to perform FFT-like algorithms over decision diagrams. This relationship is exploited here in the opposite way, the steps of FFT-like algorithms define decomposition rules for a class of decision diagrams which we call the Fourier decision diagrams. Thus, the Fourier transforms on G_i define mappings used to associate a function to the Fourier decision tree. The labels for edges of these diagrams are determined from the inverse Fourier transforms on G_i as it follows from spectral interpretation of decision diagrams [25].

Definition 4.2 *Denote by* $\mathbf{S}_f$ *the Fourier transform coefficients on a finite non-Abelian group G representable in the form (2.4). These coefficients are determined by using the relation (2.9). From (2.10), a Fourier decision tree on G is defined as the decision tree whose nodes a the i-th level represent functions*

$$f(x_i) = \sum_{w_i \in \Gamma_i} Tr(\mathbf{S}_f(w_i)\mathbf{R}_{w_i}(x_i)), \quad x_i \in G_i,$$

where Γ_i is the dual object of G_i, and $\mathbf{R}_{w_i}$ are the unitary irreducible representations of G_i.

Clearly, the Fourier decision trees on Abelian groups are derived from these on non-Abelian groups as all the unitary irreducible representations are one-dimensional, i.e., reduce to the group characters.

With this definition, Fourier decision trees are decision trees on G in which each path from the root node up to the constant nodes corresponds to a unitary irreducible representation of G. A node at the i-th level has K_i outgoing edges denoted by $\mathbf{R}_{w_i}(u), u \in \Gamma_i$. In the figures of decision trees we use the short notation w_i^j for $\mathbf{R}_{w_i}(j)$, or simply j where the context does not permit misunderstandings.

For a given f, the values of constant nodes are the Fourier coefficients of f. In that way, we perform the inverse Fourier transform in the determination of f represented by a given Fourier decision tree by using the rule in Definition 4.2 and by following the labels at the edges in the decision tree. The same can be done with any spectral transform decision diagram (STDD) [27].

The following example illustrated Definition 4.2.

Example 4.1 *Consider a function f defined on the group $G_{12} = C_2 \times S_3$ over the field $GF(11)$. Table 4.1 defines the values for f. The Fourier transforms on C_2 and S_3 over $GF(11)$ are defined by the transform matrices*

$$\mathbf{W}^{-1}(1) = 6 \begin{bmatrix} 1 & 1 \\ 1 & 10 \end{bmatrix}$$

and

$$[\mathbf{S}_3]^{-1} = 2 \begin{bmatrix} 1 & 1 & 1 & 1 & 1 & 1 \\ 1 & 1 & 1 & 10 & 10 & 10 \\ 2I & 2B & 2A & 2C & 2D & 2E \end{bmatrix}.$$

Therefore, the Fourier transform on $G_{12} = C_2 \times S_3$ over $GF(11)$ is defined as the Kronecker product of these matrices,

$$[\mathbf{R}]^{-1} = \begin{bmatrix} 1 & 1 & 1 & 1 & 1 & 1 & 1 & 1 & 1 & 1 & 1 & 1 \\ 1 & 1 & 1 & 10 & 10 & 10 & 1 & 1 & 1 & 10 & 10 & 10 \\ 2I & 2B & 2A & 2C & 2D & 2E & 2I & 2B & 2A & 2C & 2D & 2E \\ 1 & 1 & 1 & 1 & 1 & 1 & 10 & 10 & 10 & 10 & 10 & 10 \\ 1 & 1 & 1 & 10 & 10 & 10 & 10 & 10 & 10 & 1 & 1 & 1 \\ 2I & 2B & 2A & 2C & 2D & 2E & 9I & 9B & 9A & 9C & 9D & 9E \end{bmatrix}$$

The inverse transform matrices on C_2 and S_3 are

$$\mathbf{W}(1) = \begin{bmatrix} 1 & 1 \\ 1 & 10 \end{bmatrix},$$

and

$$[\mathbf{S}_3] = \begin{bmatrix} 1 & 1 & I \\ 1 & 1 & A \\ 1 & 1 & B \\ 1 & 10 & C \\ 1 & 10 & D \\ 1 & 10 & E \end{bmatrix}$$

as it follows from Table 2.5 of group representations for S_3 over $GF(11)$.

The inverse transform to reconstruct a function on G_{12} from its spectrum is given by the matrix that is the Kronecker product of these two matrices

$$[\mathbf{R}] = \mathbf{W}(1) \otimes [\mathbf{S}_3]$$

$$
= \begin{bmatrix}
1 & 1 & \mathbf{I} & 1 & 1 & \mathbf{I} \\
1 & 1 & \mathbf{A} & 1 & 1 & \mathbf{A} \\
1 & 1 & \mathbf{B} & 1 & 1 & \mathbf{B} \\
1 & 10 & \mathbf{C} & 1 & 10 & \mathbf{C} \\
1 & 10 & \mathbf{D} & 1 & 10 & \mathbf{D} \\
1 & 10 & \mathbf{E} & 1 & 10 & \mathbf{E} \\
1 & 1 & \mathbf{I} & 10 & 10 & 10\mathbf{I} \\
1 & 1 & \mathbf{A} & 10 & 10 & 10\mathbf{A} \\
1 & 1 & \mathbf{B} & 10 & 10 & 10\mathbf{B} \\
1 & 10 & \mathbf{C} & 10 & 1 & 10\mathbf{C} \\
1 & 10 & \mathbf{D} & 10 & 1 & 10\mathbf{D} \\
1 & 10 & \mathbf{E} & 10 & 1 & 10\mathbf{E}
\end{bmatrix} .
$$

The spectrum of the function f in Table 4.1 is

$$
[\mathbf{S}_f] = \begin{bmatrix}
4 \\
9 \\
2 \begin{bmatrix} 1 & 0 \\ 0 & 1 \end{bmatrix} \\
0 \\
2 \\
\begin{bmatrix} 0 & 1 \\ 1 & 4 \end{bmatrix}
\end{bmatrix} .
$$

The spectral coefficients are the values of constant nodes in a Fourier decision diagram for f. The labels at the edges are determined from the matrices $\mathbf{W}(1)$ and $[\mathbf{S}_3]$. The matrix $[\mathbf{R}]$ has a block structure that may be expressed as

$$
[\mathbf{R}] = \begin{bmatrix} [\mathbf{S}_3] & [\mathbf{S}_3] \\ [\mathbf{S}_3] & 10[\mathbf{S}_3] \end{bmatrix} .
$$

This structure is due to the Kronecker product and originates in the decomposition of G_{12}. It determines the structure of the Fourier decision trees for functions on G_{12}.
  Figure 4.12 shows the Fourier decision tree for f. This tree represents f as

$$
\begin{aligned}
f(x_1, x_2) &= 4Tr(\mathbf{R}_0(x_2)) + 9Tr(\mathbf{R}_1(x_2)) + 2Tr\left(\mathbf{R}_2(x_2) \begin{bmatrix} 1 & 0 \\ 0 & 1 \end{bmatrix}\right) \\
&\quad + (1 + 9x_1)\left(2Tr(\mathbf{R}_1(x_2)) + Tr\left(\mathbf{R}_2(x_2) \begin{bmatrix} 0 & 1 \\ 1 & 4 \end{bmatrix}\right)\right).
\end{aligned}
$$

The tree consists of two subtrees corresponding to the left and the right half of the matrix $[\mathbf{R}]^{-1}$. At each level, the transform with respect to the variable assigned to the level is performed. The left FNA_3 node corresponds to the calculation specified by the submatrix $[\mathbf{S}_3]^{-1}$ in the upper part of left half of $[\mathbf{R}]^{-1}$. When multiplied by the label 1 at the left outgoing edge of the node S_2, the calculations are repeated to correspond to the lower part of the left half of $[\mathbf{R}]^{-1}$. Similarly, the right FNA_3 node corresponds to the upper submatrix $[\mathbf{S}_3]^{-1}$ in the right half of $[\mathbf{R}]^{-1}$. Notice

Table 4.1 Truth-table for f in Example 4.1.

	$x_1 x_2$	f
0.	00	1
1.	01	0
2.	02	0
3.	03	0
4.	04	0
5.	05	1
6.	10	0
7.	11	0
8.	12	0
9.	13	1
10.	14	1
11.	15	0

that this part corresponds to $x_1 = 0$, which results in the value 1 as the label at the right outgoing edge of the node S_2. For $x_1 = 1$, this label has the value 10. When multiplied by the value 10, the right FNA_3 node performs calculations specified in the lower part of the right half of $[\mathbf{R}]^{-1}$. In this way, the Kronecker product structure of $[\mathbf{R}]^{-1}$ which determines steps of the FFT, is mapped to the recursive structure of the decision tree. It follows that to calculate values of constant nodes in the Fourier decision tree, we perform the Fourier transform. Conversely, through labels at the edges, we perform the inverse Fourier transform to read values of the function represented as stated in Definition 4.2, and calculations are performed as FFT over the decision tree.

As is noted above, the Fourier coefficients corresponding to the group representations with orders $r_w > 1$, $w = 1, \ldots K - 1$, are $(r_w \times r_w)$ matrices. Therefore, Fourier decision trees on finite non-Abelian groups are matrix-valued decision trees. The matrix-valued coefficients may be also represented by the decision trees by using the method of representation of matrices by the decision diagrams [5]. In that way, we derive the integer-valued or complex-valued Fourier decision trees on finite non-Abelian groups, depending on the values of $\{R_w^{(i,j)}\}$ that are the values of constant nodes in these decision trees. In that case, the number of constant nodes in a integer or complex-valued Fourier decision trees on a non-Abelian group is equal to that in Fourier or other decision trees on an Abelian group of the same order. Comparing to the matrix-valued Fourier decision trees, the number of levels in an integer-valued or complex-valued Fourier decision tree is increased to represent the matrix-valued constant nodes. However, some non-terminal nodes still may be saved, since not all the Fourier coefficients are matrix-valued.

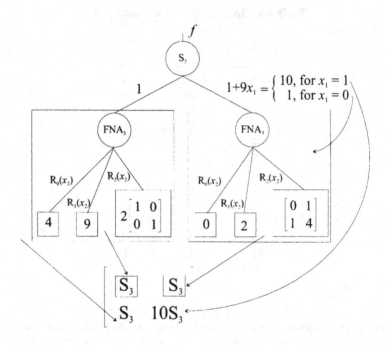

Fig. 4.12 Fourier decision diagram for f in Example 4.1.

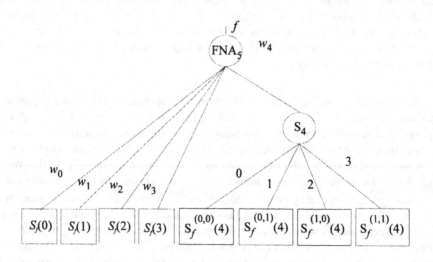

Fig. 4.13 Complex-valued FNADD for Q_2.

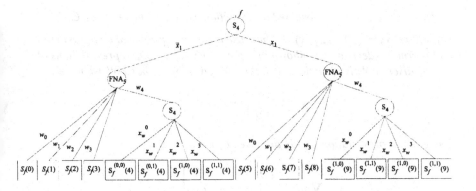

Fig. 4.14 Decision tree with the Shannon node S_2 and FNADD nodes for Q_2.

The non-terminal nodes in Fourier decision trees are denoted by FA_i for Abelian and FNA_i for non-Abelian groups. The index i denotes the number of outgoing edges and it is equal to the cardinality K_i of the dual object Γ_i of G_i.

The following example illustrates the effect of the decomposition of the domain group of f into non-Abelian subgroups.

Example 4.2 *In a Shannon tree, a subtree of order 8 (Figure 4.6(a) and Figure 4.8) may be replaced by the Fourier subtree on Q_2 shown in Figure 4.13. Compared to the subtree in Figure 4.6(a), we have two levels and two non terminal nodes, but we keep the reduction possibility of order 4 for S_4 nodes and 5 for FNA_5 nodes. Compared to the subtree in Figure 4.8, we have a saving of one non-terminal node.*

Figure 4.14 shows the Fourier decision tree on $G_{16} = C_2 \times Q_2$. Compared to the Shannon tree in Figure 4.1, we have reduction of both number of levels and total of nodes. Compared to the QDT in Figure 4.4, we have one level more, but reduced the number of nodes per levels. The number of nodes is increased, but the reduction possibilities and the number of outgoing edges per nodes are much better. The number of nodes per levels equals that in Figure 4.7. Compared to the decision tree in Figure 4.9, we have the same number of levels, but reduced number of non-terminal nodes. Thus, the Fourier DT in Figure 4.14 is a compromise between those in Figure 4.1 and Figure 4.4 and Figure 4.9 with respect to the number of nodes and number of nodes per level. Taking into account the reduction possibilities in a DT, Fourier decision diagrams may offer more efficient representations that MTDDs.

The following example illustrates the use of Fourier decision trees on both Abelian and non-Abelian groups in representation of discrete functions.

Example 4.3 *A function f defined on a set of six points can be represented by the truth-vector $\mathbf{F} = [f(0), f(1), f(2), f(3), f(4), f(5)]^T$. In a direct graphical representation $\mathbf{F}$ is shown in Figure 4.15. This figure corresponds to the truth-vector representation of f over the cyclic group of order 6.*

This function can be considered as a function $f(x_1, x_2)$ on the group $G_6 = C_2 \times G_3$, where $G_3 = (\{0, 1, 2\}, \overset{\oplus}{3})$. In that case it can be represented by the generalized multi-terminal decision tree shown in Figure 4.16. The group representations of C_2 are the discrete Walsh functions and, thus are given by the basic Walsh matrix

$$\mathbf{W}(1) = \frac{1}{2} \begin{bmatrix} 1 & 1 \\ 1 & -1 \end{bmatrix}.$$

The group representations of G_3 are given by the matrix

$$\mathbf{V}(1) = \frac{1}{3} \begin{bmatrix} 1 & 1 & 1 \\ 1 & e_2 & e_1 \\ 1 & e_1 & e_2 \end{bmatrix},$$

where $e_1 = -\frac{1}{2}(1 - i\sqrt{3})$ and $e_2 = -\frac{1}{2}(1 + i\sqrt{3})$. Therefore, the Fourier transform on the Abelian group $G_6 = C_2 \times G_3$ is defined by the matrix

$$\mathbf{V} = (\mathbf{W}(1) \otimes \mathbf{V}(1)),$$

where $\otimes$ denotes the Kronecker product.

In that case, f can be represented by the Fourier decision tree shown in Figure 4.17. This decision tree represents function

$$\begin{aligned} f = \quad & 1 \cdot (1 \cdot S_f(00) + e_1 S_f(01) + e_2 S_f(02) \\ & -1 \cdot (1 \cdot S_f(10) + e_1 S_f(11) + e_2 S_f(12)). \end{aligned}$$

The function f can be considered as a function on the symmetric group of permutations S_3 and in that case can be represented by the Fourier decision tree shown in Figure 4.18. This tree represents the function

$$f = R_0 S_f(0) + R_1 S_f(1) + Tr(\mathbf{R}_2 \mathbf{S}_f(2)).$$

The value of the constant node corresponding to $\mathbf{S}_f(2)$ is a (2×2) matrix. Thus, this tree is a matrix-valued tree. If matrix-valued constant nodes are not allowed, $\mathbf{S}_f(2)$ may be also represented by a tree. For that reason, we write $\mathbf{S}_f(2)$ as a function q on the cyclic group G_4 and represent it by the truth-vector $\mathbf{Q} = [S_f^{(0,0)}, S_f^{(0,1)}, S_f^{(1,0)}, S_f^{(1,1)}]^T$. In that way, $\mathbf{S}_2$ can be represented by the MTBDDs on G_4. The corresponding Fourier decision tree is a complex-valued decision tree. This tree is shown in Figure 4.19. From this figure, even with complex-valued Fourier decision tree, we still have savings in non-terminal nodes comparing to the decision trees based on Abelian groups (Figure 4.16 and 4.17). We also have further possibilities for the reduction of this tree. Depending on the values of elements of $\mathbf{S}_f(2)$, we can use the Fourier decision tree on G_4 to represent it and to save eventually some of the nodes representing elements of $\mathbf{S}_f(2)$. The corresponding decision trees representation of f is as in Figure 4.20. The values of constant nodes are the Fourier coefficients S_q of q on G_4.

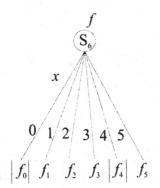

Fig. 4.15 One-level multi-terminal decision tree of f in Example 4.3 on G_6.

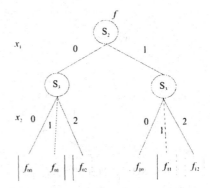

Fig. 4.16 Two-level multi-terminal decision tree of f in Example 4.3 on $G_6 = C_2 \times G_3$.

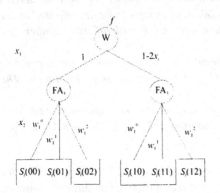

Fig. 4.17 Two-level Fourier decision tree of f in Example 4.3 on the Abelian group $G_6 = C_2 \times G_3$.

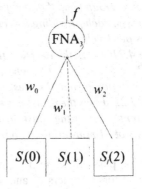

Fig. 4.18 One-level Fourier decision tree of f in Example 4.3 on non-Abelian group S_3.

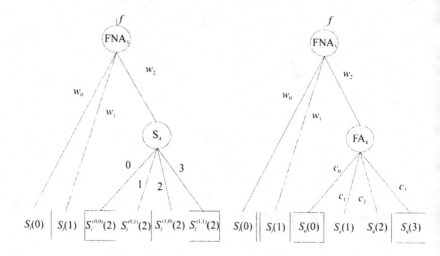

Fig. 4.19 Complex-valued Fourier DT on S_3 with MTBDD for $S_f(2)$.

Fig. 4.20 Complex-valued Fourier DT on S_3 with Fourier DT for $S_f(2)$.

The following example further elaborates the use of non-Abelian groups in optimization of decision diagrams.

Example 4.4 *A function f defined on a set of 36 points can be considered as a function on a group G_{36} of order $g = 36$. This group can be decomposed in several different ways, each providing a decision tree on G_{36} with different number of levels and nodes per levels.*

Figure 4.21 shows the decision tree on G_{36} corresponding to the decomposition $G_{36} = C_2 \times C_2 \times C_3 \times C_3$, where C_i denotes a cyclic group of order i. The DT consists of 19 non-terminal nodes.

Figure 4.22 shows the decision tree corresponding to the same decomposition, but with different order of the constituent subgroups $G_{36} = C_3 \times C_3 \times C_2 \times C_2$. This DT consists of 31 non-terminal nodes.

Reduction possibilities in decision trees in Figure 4.21 and Figure 4.22 depend on the subvectors of orders 2^k and 3^k for nodes with two and three outgoing edges, respectively.

Fourier decision trees on finite non-Abelian groups (FNADDs) take advantage of the matrix notation of group representations and, therefore, the spectral coefficients. The cardinality of the dual object Γ of a non-Abelian group G is always smaller than the order g of G. The same is true if the cardinalities of the dual objects Γ_i and the orders of the constituent subgroups G_i are compared. Note that in the matrix notation the group representations of G are generated through the generalized Kronecker product of the group representations of the constituent subgroups G_i. Therefore, the

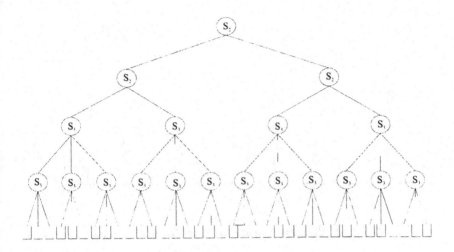

Fig. 4.21 Decision tree on $G_{36} = C_2 \times C_2 \times C_3 \times C_3$.

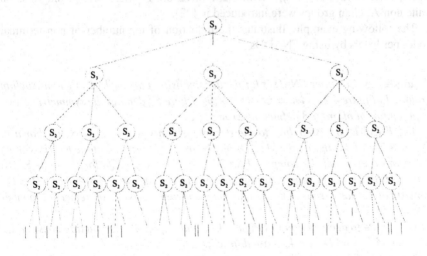

Fig. 4.22 Decision tree on $G_{36} = C_3 \times C_3 \times C_2 \times C_2$.

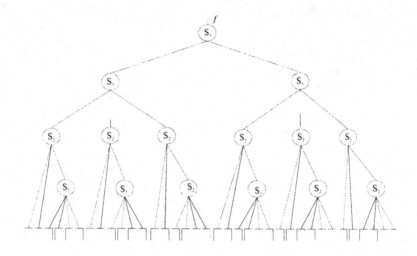

Fig. 4.23   Complex-valued FNADT on $G_{36} = C_2 \times C_3 \times S_3$.

number of nodes at intermediate levels is determined by the cardinality K_i of Γ_i. It follows that some non-terminal nodes may be saved. In that way, FNADDs permit the reduction of both the number of levels and the number of nodes per levels in a decision tree. Unlike decision trees on Abelian groups, in FNADDs the use of large subgroups to reduce the number of levels does not increase the number of nodes per levels, since always $K_i < g_i$. With that motivation, Fourier decision diagrams on finite non-Abelian groups were introduced in [22].

The following example illustrates the reduction of the number of non-terminal nodes per levels by using FNADDs.

Example 4.5 *Consider FNADTs for the decomposition of G_{36} by using non-Abelian groups. In Figure 4.23, $G_{36} = C_2 \times C_3 \times S_3$, where S_3 denotes the symmetric group of permutation of order 3 defined in Example 2.2.*

The FNADT corresponding to the decomposition $G_{36} = S_3 \times S_3$ is shown in Figure 4.24. In this figure, the MTDDs on G_4 are used to represent the matrix-valued nodes of order 2 and the complex-FNADT on $G_{16} = C_2 \times Q_2$ for the node $\mathbf{S}_f(8)$ with MTDTs on G_4 to represent the matrix-valued nodes in this sub-tree. This DT consists of 13 non-terminal nodes and the disadvantage is the incremented number of levels to 6 instead of 5.

Compared to decision trees in Figure 4.21 and Figure 4.22, reduction possibilities in Figure 4.23 and Figure 4.24 are different.

Example 4.5 shows that by using decision trees on non-Abelian groups, the number of non-terminal nodes may be drastically reduced without essentially decreasing reduction possibilities in the decision tree.

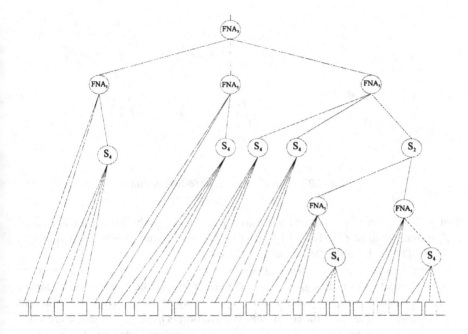

Fig. 4.24 Complex-valued FNADT on $G_{36} = S_3 \times S_3$.

4.3.2 Fourier decision diagrams

Fourier decision diagrams are derived by reducing Fourier decision trees. The reduction is done by using the generalized BDD reduction rules introduced for the reduction of WDDs [27].

Definition 4.3 *(Generalized BDD reduction rules)*

1. *Delete all the redundant nodes where both edges point to the same node and connect the incoming edges of the deleted nodes to the corresponding successors. Relabel these edges as is shown in Figure 4.25(a).*

2. *Share all the equivalent sub-graphs, Figure 4.25(b).*

Definition 4.4 *Fourier decision diagrams are decision diagrams derived from the Fourier decision trees by using the generalized BDD reduction rules. A Fourier decision diagram is reduced if further reduction with the same rules is impossible.*

The discussion in Section 4.1 about the reduction possibilities applies to Fourier decision trees.

The main difference with MTDDs is that the reduction possibilities in the Fourier decision tree depend on the relationships among the values of Fourier coefficients S_f

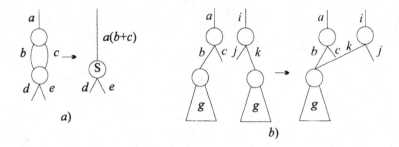

Fig. 4.25 Generalized BDD reduction rules.

of f. The order of the reduction possibility is determined by the cardinality K_i of Γ_i instead of g_i of G_i. Thus, in Fourier decision diagrams on non-Abelian groups (FNADDs) we have smaller reduction possibilities.

If for a given assignment of values of a variable $x_i = p_i, p_i \in \{0, \ldots, g_i\}$, there are equal subvectors of orders $g_k, k < i$, in the vector $[\mathbf{S}_f]$ of the Fourier coefficients of f, then the corresponding nodes at the k-th level may be joined. Thus, the redundant nodes may be deleted. If the equal subvectors of orders $g_k, k < i$, in $[\mathbf{S}_f]$ correspond to different assignments of values of x_i, the corresponding nodes may be shared.

4.4 DISCUSSION OF DIFFERENT DECOMPOSITIONS

To estimate the complexity of a decision diagram representation for a given function f, the size of a decision diagram is defined as the number of nodes in the decision diagram.

In decision diagrams based calculation procedures, some calculations should be performed at each non-terminal node. The time complexity of the calculation procedure is, therefore, expressed as the size of decision diagram times the processing time per node. In some procedures where the processing of nodes differ, for example for nodes at different levels, we use the maximal or possibly average time per node.

However, two additional parameters are important if we consider the complexity of functions realizations derived from their decision diagram representations:

1. area that the produced network occupies,

2. propagation delay.

Therefore, we should speak about the space-time complexity. In decision diagram based realizations, the time complexity depends on the number of levels in the decision diagram. To estimate the space complexity, we can use the product of the number of levels and the number of nodes per level (the maximal number for the worst case and the average number of nodes per level otherwise).

Therefore, the relevant parameters to estimate the space-time complexity of decision diagram based realizations for a given function f are

Table 4.2 Space-time complexity of DTs on G_{16}.

Decomposition	Space-time complexity
$C_2 \times C_2 \times C_2 \times C_2$	$2 \cdot 2 \cdot 2 \cdot 2$
$C_4 \times C_4$	$4 \cdot 4$
$C_2 \times C_8$	$2 \cdot 8$
$C_2 \times C_2 \times C_4$	$2 \cdot 2 \cdot 4$
$Q_2 \times Q_2$	$2 \cdot (4 + 1 \cdot 4) = 2 \cdot 4 + 2 \cdot 1 \cdot 4$

1. number of levels,

2. maximal number of nodes per level,

3. average number of nodes per level.

Example 4.9 compares these parameters for the 3-bit multiplier representation by MTDDs and Fourier decision diagrams for different decompositions of the domain group.

However, to study the effect of different decoding of variables, or in the group-theoretic interpretation, the effect of different decompositions of G, we normalized the space-time complexity with respect to the order of G. In this way, the decision trees in Figure 4.21, 4.22, 4.23 and 4.24 have all the normalized space-time complexity equal 16, since they all represent functions whose truth-vectors are of order 16. We represent the time complexity (expressed through the number of levels) by the number of factors in the decomposition of 16. We represent the space complexity of a level by the number of outgoing edges of the nodes the level consists of. Table 4.2 shows the space-time complexity of decision trees for different decompositions of G_{16}. In the first three decompositions we can speak about the multiplicative time complexities of orders 4, 2 and 2, respectively (number of factors). The space complexities are 2, 4 and 8. With the Fourier decision diagram we have the decomposition of the complexity into two additive parts. The time complexities of these parts are 2 and 3, respectively. The space complexities of them are 2, 5 and 4, that is simpler than in the other cases where the time complexity is equal to 2. The decomposition of the space-time complexity in the Fourier decision diagram into two additive parts explains that we can save a non-terminal node in the replacement of the subtree of order 8 in Figure 4.6(a) and Figure 4.8 by the Fourier decision diagram on Q_2.

Combination of Abelian and non-Abelian groups of different orders provides decision diagrams of different complexity with respect to both number of levels and nodes. A suitable combination can provide an optimal representation with respect to both criteria.

FNADDs permit a two-step optimization. First, we can use large non-Abelian subgroups G_i to define a decision tree with a required number of levels and to reduce the number of non-terminal nodes per levels at the same time. Then, we can represent the matrix-valued constant nodes by decision trees on Abelian groups and do the optimization by choosing among various possible decision trees on different Abelian groups.

4.4.1 Algorithm for optimization of DDs

From the above consideration the following algorithm for the optimization of decision diagrams representations of discrete functions may be formulated.

Algorithm for optimization of FNADDs representations

1. *Given a function f defined on a set D of d elements. Assume for D the structure of a group G of order $g = d$.*

2. *Determine the required number n of levels in the decision diagrams representation of f. Determine possible decompositions of G into n subgroups G_i.*

3. *Try different group structures for each G_i by using the non-Abelian groups for the largest G_i's. Draw the matrix-valued FNADDs for each possible combination of subgroups and calculate the number of non-terminal nodes and constant nodes. Chose the smallest size DD with the most planar structure.*

4. *End of the algorithm, if matrix-valued FNADDs are allowed. Otherwise, try representations of matrix-valued constant nodes by corresponding generalizations of MTBDDs and FNADDs. Chose the optimal representation for each node.*

To reduce the number of possible combinations of various decompositions of G and to get a quick solution, it is a reasonable to use decomposition into the subgroups of increasing orders. This view is based upon the previous experience in the closely related problem of efficient calculation of the Fourier transform on finite non-Abelian groups [9], [12].

4.5 REPRESENTATION OF TWO-VARIABLE FUNCTION GENERATOR

The representation of the function f given by the truth-vector $\mathbf{F} = [0, 1, 2, 3, 4, 5, 6, 7]^T$ was used in [13] as an example to illustrate the efficiency of Edge-valued Binary Decision diagrams (EVBDDs). We will use quite a similar example to illustrate the efficiency of FNADDs.

Notice that EVBDDs are decision diagrams with attributed edges [25], which represent functions in the form of the arithmetic polynomials [21]. Arithmetic spectral transform decision diagrams (ACDDs) [27] do not have attributed edges, the same as FNADDs, but also represent functions in the form of arithmetic polynomials. Therefore, we will provide comparisons with ACDDs. Notice that definitions of the Arithmetic transform and arithmetic polynomials are considered in Section 5.3, and

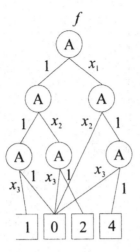

Fig. 4.26 Arithmetic transform decision diagram for f in the Example 4.6.

here we briefly recall the definition of the Arithmetic spectral transform decision diagrams.

Definition 4.5 *[27] Arithmetic spectral transform decision diagrams (ACDDs) are binary decision diagrams, i.e., consists of nodes with two outgoing edges, where the values of constant nodes are the arithmetic spectral coefficients, and labels at the edges are 1 and x_i.*

Example 4.6 *For a function f given by the vector $\mathbf{F} = [0, 1, 2, 4, 5, 6, 7]^T$, the arithmetic spectrum is calculated by using the arithmetic transform matrix for $n = 3$ defined in Example 5.3, thus, it is*

$$\begin{aligned} \mathbf{A}_f &= \mathbf{A}(3)\mathbf{F} \\ &= [0, 1, 2, 0, 4, 0, 0, 0]^T. \end{aligned}$$

Figure 4.26 shows the ACDD for f.

Example 4.7 *Realization of switching functions by logic networks containing generators of all switching functions as basic design modules has been discussed already by Shannon [20].*

A two-variable functions generator is efficiently applied as a design module also in present technologies, see for example [14], [15]. Figure 4.27 and Figure 4.28 show two-variable function generators in the notation by Shannon and with more recent symbols for elementary logic circuits [15].

The outputs of the two-variable function generator is described by the set of switching functions $\{xy, x\bar{y}, x, \bar{x}y, y, x \oplus y, x \vee y, \bar{x} \cdot \bar{y}, \bar{x} \oplus y, \bar{y}, x \vee \bar{y}, \bar{x}, \bar{x} \vee y, \bar{x} \vee \bar{y}\}$. We

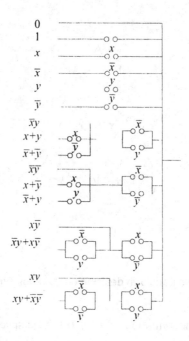

Fig. 4.27 Two-variable function generator in Shannon notations.

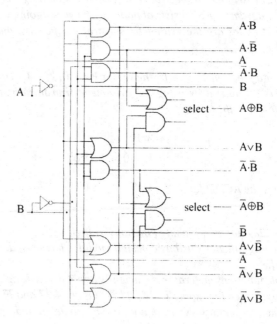

Fig. 4.28 Two-variable function generator.

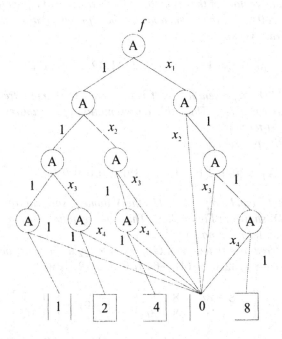

Fig. 4.29 ACDD representation of $f(z)$ for TVFG.

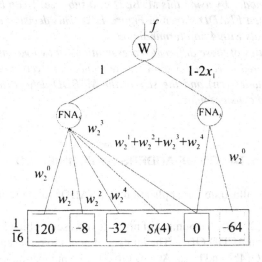

Fig. 4.30 Matrix-valued FNADD representation of $f(z)$ for TVFG.

represent the truth-vectors of these functions by their decimal equivalents. Together with the constants 0 and 1, these functions can be represented as an integer function f given by the truth-vector

$$\mathbf{F} = [0, 1, 2, 3, 4, 5, 6, 7, 8, 9, 10, 11, 12, 13, 14, 15]^T.$$

Therefore, a MTBDD representing this f is the complete Shannon tree for switching functions of four variables. Thus, this representation of f requires 31 nodes in 5 levels with 15 non-terminal nodes.

   The arithmetic spectrum of f is

$$\mathbf{A}_f = [0, 1, 2, 0, 4, 0, 0, 0, 8, 0, 0, 0, 0, 0, 0, 0]^T,$$

and it follows that f can be represented by the Arithmetic spectral transform decision diagram (ACDD) shown in Figure 4.29. This representation requires 15 nodes in 5 levels with 10 non-terminal nodes.

   If f is considered as a function on $G_{16} = C_2 \times Q_2$ its Fourier spectrum is given by $\mathbf{S}_f = \frac{1}{16}[120, -8, -32, 0, \mathbf{S}_f(4), -64, 0, 0, 0, \mathbf{S}_f(9)]^T$, with

$$\mathbf{S}_f(4) = \begin{bmatrix} -8+8i & -8-8i \\ 8-8i & -8-8i \end{bmatrix}, \quad \mathbf{S}_f(9) = \begin{bmatrix} 0 & 0 \\ 0 & 0 \end{bmatrix}.$$

Therefore, f can be represented by the matrix-valued FNADD in Figure 4.30. The representation requires 9 nodes in 3 levels with 3 non-terminal nodes.

   The matrix-valued node $\mathbf{S}_f(4)$ can be represented by the MTBDD on G_4 shown in Figure 4.31. The impact of the zero matrix for $\mathbf{S}_f(9)$ is equal to the ordinary zero-valued constant node. By using this MTBDD as a sub-tree, f can be represented by the complex-valued FNADD shown in Figure 4.32. This decision diagram requires 12 nodes in 4 levels with 4 non-terminal nodes.

   The complexities of these different representations of two-variable function generator are compared in Table 4.3 showing the number of levels, non-terminal nodes (ntn), constant nodes (cn) and the size of the MTBDD defined as the number of non-terminal and constant nodes.

4.6   REPRESENTATION OF ADDERS BY FOURIER DD

In this section we illustrate the application of FNADDs to the representation of an n-bit adder.

   It is shown in [27] that the number of nodes to represent n-bit adder by a MTBDD is $O(2^n)$ and by an ACDD $O(n)$. The number of nodes to represent n-bit multiplier by a MTBDD is $O(4^n)$ and by an ACDD $O(n^2)$. The number of levels in MTBDDs and ACDDs is always equal to n.

   It is shown on the example of 2-bit adders and multipliers that FNADDs permit some considerable savings in both number of levels and non-terminal nodes. Besides

Table 4.3 Complexity of representation of TVFG in terms of the number of levels, non-terminal nodes (ntn), constant nodes (cn) and sizes.

Number of	levels	ntn	cn	size	
MTBDD	5	15	16	31	
ACDD	5	10	5	15	Figure 4.29
Matrix-valued FNADD	3	3	6	9	Figure 4.30
Complex-valued FNADD	4	4	8	12	Figure 4.32

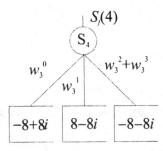

Fig. 4.31 MTBDD representation of $S_f(4)$ for TVFG.

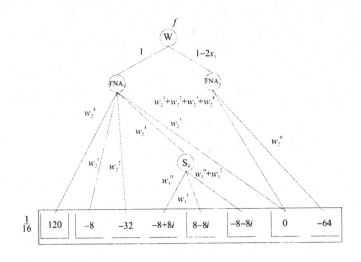

Fig. 4.32 Complex-valued FNADD representation of f for TVFG.

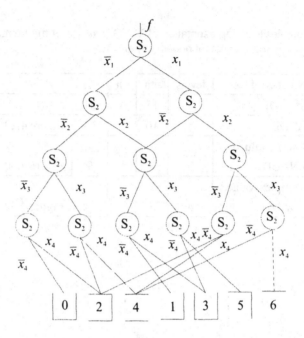

Fig. 4.33   MTBDD representation of f for 2-bit adder.

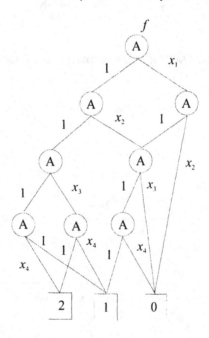

Fig. 4.34   ACDD representation of f for 2-bit adder.

that, the savings in the total number of nodes compared to MTBDDs representations are achieved even with complex-valued FNADDs. In that case the total number of nodes is equal to that used with ACDDs.

Example 4.8 *The outputs of the 2-bit adder may be represented by the integer function* f *given by the truth-vector* $\mathbf{F} = [0, 2, 2, 4, 1, 3, 3, 5, 1, 3, 3, 5, 2, 4, 4, 6]^T$ *[14].* *Therefore, it may be represented by the MTBDD shown in Figure 4.33. The representation requires 19 nodes over 5 levels.*

The arithmetic spectrum of f *is* $\mathbf{A}_f = [0, 2, 2, 0, 1, 0, 0, 0, 1, 0, 0, 0, 0, 0, 0, 0]^T$. *Therefore,* f *may be represented by ACDD shown in Figure 4.34. The representation requires 11 nodes in 5 levels.*

If f *is considered as a function on the finite non-Abelian group* $G_{16} = C_2 \times Q_2$, *where* C_2 *is the basic dyadic group and* Q_2 *is the quaternion group, the Fourier spectrum of* f *is*

$$\mathbf{S}_f = \frac{1}{16} [6, -2, -1, 0, \mathbf{S}_f(4), -1, 0, 0, 0, \mathbf{S}_f(9)]$$

where $\mathbf{S}_f(4) = \begin{bmatrix} -1+i & -1-i \\ 1-i & -1-i \end{bmatrix}$ *and* $\mathbf{S}_f(9) = \begin{bmatrix} 0 & 0 \\ 0 & 0 \end{bmatrix}$.

Therefore, it may be represented by the matrix-valued Fourier decision tree shown in Figure 4.35, that can be reduced into the matrix-valued Fourier decision diagrams shown in Figure 4.36. The representation requires 8 nodes in 3 levels.

The value of the constant nodes $\mathbf{S}_f(4)$ *and* $\mathbf{S}_f(9)$ *are* (2×2) *matrices. The impact of the zero matrix for* $\mathbf{S}_f(9)$ *is equal to the ordinary zero-valued constant node.*

If matrix-valued constant nodes are not allowed, that matrix can be considered as the function $\mathbf{S}_f = [-1+i, 1-i, -1-i, -1-i]^T$. *If this function* $\mathbf{S}_f$ *is considered as a function on the cyclic group of order 4,* G_4, *it can be represented by the quaternary decision diagram QDD [18] shown in Figure 4.37. By using this decision diagram, the complex-valued FNADD representing* f *for 2-bit adder may be derived as in Figure 4.38.*

The complexities of representing the 2-bit adder with these various decision diagrams are compared in Table 4.4, where number of levels, non-terminal nodes (ntn), constant nodes (cn), and sizes of decision diagrams are shown.

4.7   REPRESENTATION OF MULTIPLIERS BY FOURIER DD

The above method is applied to the optimization of the representation of multipliers by Fourier decision diagrams. For simplicity, the presentation is given on the examples of 2-bit and 3-bit multipliers. Similar conclusions may be derived in the general case.

The following example considers the efficiency of representing the 2-bit multiplier by different decision diagrams.

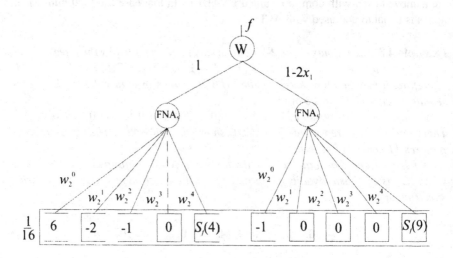

Fig. 4.35 Fourier DT of f for 2-bit adder.

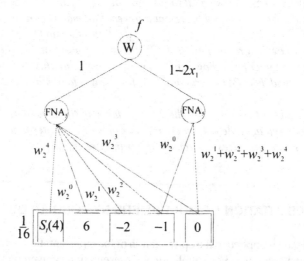

Fig. 4.36 Matrix-valued FNADD of f for 2-bit adder.

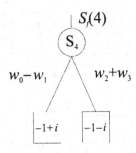

Fig. 4.37 MTBDD representation of $S_f(4)$ for 2-bit adder.

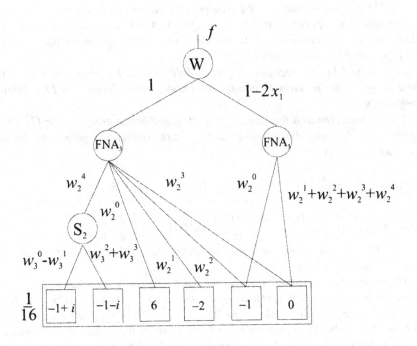

Fig. 4.38 Complex-valued FNADD of f for 2-bit adder.

Table 4.4 Complexities of representing the 2-bit adder in terms of levels, number of non-terminal nodes (ntn), constant nodes (cn), and sizes.

Number of	levels	ntn	cn	size	
MTBDD	5	12	7	19	Figure 4.33
ACDD	5	8	3	11	Figure 4.34
Matrix-valued FNADD	3	3	5	8	Figure 4.36
Complex-valued FNADD	4	4	6	10	Figure 4.38

Example 4.9 *The outputs of the 2-bit multiplier may be represented by the integer function f given by the truth-vector [14]*

$$\mathbf{F} = [0,0,0,4,0,0,2,6,0,2,0,6,1,3,3,9]^T.$$

The MTBDD representation of f requires 20 nodes over 5 levels.

The arithmetic spectrum of f is $\mathbf{A}_f = [0,0,0,4,0,0,2,0,0,2,0,0,1,0,0,0]^T$. *Therefore f may be represented by an arithmetic spectral transform decision diagrams (ACDD) with 15 nodes in 5 levels.*

Edge-valued binary decision diagrams (EVBDD) [13] representation of f for the 2-bit multiplier requires 10 non-terminal nodes over 5 levels with 11 weighting coefficients.

If f is considered as a function on the finite non-Abelian group $G_{16} = C_2 \times Q_2$, *where* C_2 *is the basic dyadic group and* Q_2 *is the quaternion group, the Fourier spectrum of* $f(z)$ *is*

$$\mathbf{S}_f = \frac{1}{16} \left[36, -24, -12, 0, \mathbf{S}_f(4), -12, 8, 4, 0, \mathbf{S}_f(9) \right]^T$$

where $\mathbf{S}_f(4) = \begin{bmatrix} 16i & -8-24i \\ 8-24i & -16i \end{bmatrix}$ *and* $\mathbf{S}_f(9) = \begin{bmatrix} 0 & 0 \\ 0 & 0 \end{bmatrix}$.

Therefore, it may be represented by the matrix-valued Fourier decision diagram which requires 8 nodes in 3 levels.

The values of the constant nodes $\mathbf{S}_f(4)$ *and* $\mathbf{S}_f(9)$ *are* (2×2) *matrices.*

If the matrix-valued constant nodes are not allowed, that matrix can be considered as the function $\mathbf{S}_f = [16i, 8-24i, -8-24i, -16i]^T$. *If this function* $\mathbf{S}_f$ *is considered as a function on the cyclic group of order 4,* G_4, *it can be represented by a quaternary decision diagram QDD [18]. By using this decision diagram, the complex-valued FNADD may be derived. This decision diagram representation of the 2-bit multiplier requires 13 nodes in 4 levels.*

The complexities of representing the 2-bit multiplier with these various decision diagrams are compared in Table 4.5.

Table 4.5 Complexities of representing the 2-bit multiplier in terms of the number of levels, non-terminal nodes (ntn), constant nodes (cn) and sizes.

Number of	levels	ntn	cn	size
MTBDD	5	13	7	20
ACDD	5	11	4	15
EVBDD	5	10	11	21
Matrix-valued FNADD	3	3	7	10
Complex-valued FNADD	4	4	9	13

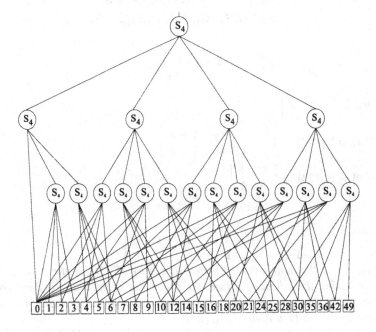

Fig. 4.39 MTDD for the 3-bit multiplier on $G_{64} = C_4 \times C_4 \times C_4$.

In the above example, the reduction of levels is achieved by using Fourier decision diagrams. The other decision diagrams on Abelian groups $G_{16} = C_2 \times C_2 \times C_2 \times C_2$ do not offer such possibility. The reduction of levels may be achieved with QDDs and their corresponding generalizations for different decompositions of the domain group G for f. The following example studies the effect of different decompositions of the domain group G_{64} in the representation of 3-bit multiplier with MTDDs and compares these representations with a Fourier decision diagram representation.

Example 4.10 *The outputs of the 3-bit multiplier may be represented by the integer function f given by the truth-vector [14]*

$$\mathbf{F} = [0,0,0,0,0,0,0,0,0,1,2,3,4,5,6,7,$$
$$0,2,4,6,8,10,12,14,0,3,6,9,12,15,18,21,$$
$$0,4,8,12,16,20,24,28,0,5,10,15,20,25,30,35,$$
$$0,6,12,18,24,30,36,42,0,7,14,21,28,35,42,49]^T.$$

If f is considered as a function on the finite non-Abelian group $G_{64} = Q_2 \times Q_2$, where Q_2 is the quaternion group, the Fourier spectrum of f is

$$\mathbf{S}_f = \frac{1}{4}[49,-7,28,0,\mathbf{S}_f(4),-7,1,4,0,\mathbf{S}_f(9)$$
$$-28,4,16,0,\mathbf{S}_f(14),0,0,0,0,\mathbf{S}_f(19)$$
$$\mathbf{S}_f(20),\mathbf{S}_f(21),\mathbf{S}_f(22),\mathbf{S}_f(23),\mathbf{S}_f(24)]^T,$$

where

$$\mathbf{S}_f(4) = \mathbf{S}_f(20) = -7\begin{bmatrix} 1-i & -1+i \\ -1+i & 1+i \end{bmatrix},$$

$$\mathbf{S}_f(9) = \mathbf{S}_f(21) = \begin{bmatrix} 1-i & -1+i \\ -1+i & 1+i \end{bmatrix},$$

$$\mathbf{S}_f(14) = \mathbf{S}_f(22) = 4\begin{bmatrix} 1-i & -1+i \\ -1+i & 1+i \end{bmatrix},$$

$$\mathbf{S}_f(19) = \mathbf{S}_f(23) = \begin{bmatrix} 0 & 0 \\ 0 & 0 \end{bmatrix}$$

$$\mathbf{S}_f(24) = 2\begin{bmatrix} -i & -1+i & 1 & -1-i \\ -1+i & 1 & 1+i & i \\ i & 1-i & 1 & -1-i \\ -1+i & -1 & 1+i & i \end{bmatrix}.$$

We represent the matrix-valued nodes simply by MTDTs of the corresponding orders by taking the advantages from the properties of the matrix-valued Fourier coefficients[1]. In that way we produce the complex-valued Fourier decision diagrams representing the 3-bit multiplier [24], [25].

The complexities of representing the 3-bit multiplier with MTDDs for different decompositions of G_{64} and complex-valued Fourier DD on $G_{64} = Q_2 \times Q_2$ are compared in Table 4.6. This table shows the number of levels, non-terminal nodes (ntn), maximum number of nodes per level (max), minimum number of nodes per level (min), average number of nodes per level (av), number of constant nodes(cn), total of nodes, i.e., the size (s), and the maximum number of edges per node (e).

Fourier decision diagrams permit the use of nodes with negated edges (FNADD n.e. in Table 4.6) [24], [25], which means that the number of constant nodes may be

---

[1]In many cases, matrix-valued Fourier coefficients are symmetric, Hermitean or skew-Hermitean matrices.

considerably reduced. In this example, however, the negative edges can not be used with MTDDs.

Figure 4.39 shows an MTDD for the 3-bit multiplier for the decomposition $G_{64} = C_4 \times C_4 \times C_4$. Figure 4.40 shows an FNADD for the 3-bit multiplier for the decomposition $G_{64} = Q_2 \times Q_2$ [24], [25].

Table 4.6 Complexities of representing the 3-bit multiplier in terms of the number of levels, non-terminal nodes (ntn), maximum (max) and minimum (min) number of nodes per level, average number per level (av), constant nodes (cn), sizes (s), and maximum number of edges per level (e).

Number of	non-terminal							
	levels	ntn	max	min	av	cn	s	e
MTBDD C_2^6	5	55	28	2	11	26	81	2
MTDD $C_2^2 \times C_4^2$	3	20	14	2	6.66	26	46	4
MTDD $C_4 \times C_4 \times C_4$	2	18	14	4	9	26	44	4
MTDD $C_8 \times C_8$	1	7	7	7	7	26	34	8
FNADD $Q_2 \times Q_2$	2	11	7	4	5.5	25	36	5
FNADD n. e. $Q_2 \times Q_2$	2	11	7	4	5.5	17	28	5

4.8   COMPLEXITY OF FNADD

In this section, we compare various decision diagrams on dyadic groups with FNADDs on the quaternion groups. Multiple-output functions are represented by Shared BDDs (SBDDs), Multi-terminal binary decision diagrams (MTBDDs) [14], and FNADDs. For representation by MTBDDs and FNADDs, the benchmark functions are first represented by the integer-valued functions f. Depending on the value of n, in FNADDs on quaternion groups $C_2 Q_2^r$ or $C_4 Q_2^r$, we use as the root node the Shannon nodes with two or four outgoing edges.

Table 4.7 compares sizes (s) and widths (w) of SBDDs and FNADDs for various benchmark functions. For FNADDs the values of non-terminal nodes (ntn) and constant nodes (cn) are shown separately. Thus, the size of FNADDs is the sum of these

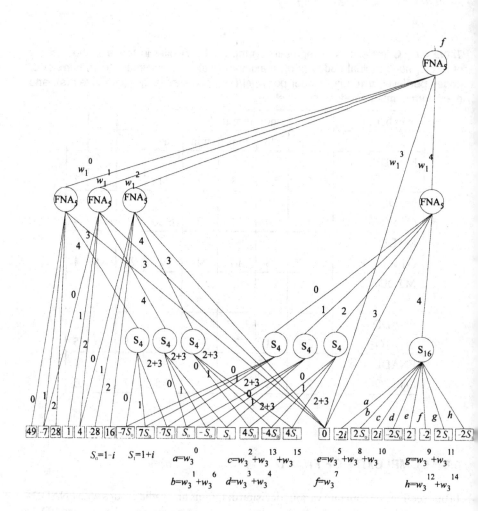

Fig. 4.40 FNADD for the 3-bit multiplier on $G_{64} = Q_2 \times Q_2$.

Table 4.7 SBDDs and FNADDs for benchmark functions.

f	SBDD			FNADD						
	s	w	a	ntn	cn	s	w	a	r_n	group
5xp1	90	25	2250	39	128	167	18	3006	0.35	$C_2Q_2^2$
bw	116	37	4292	9	25	34	4	136	0.36	C_4Q_2
con1	20	5	100	13	12	25	6	150	1.83	$C_2Q_2^2$
rd53	25	6	150	7	14	21	3	63	0.50	C_4Q_2
rd73	45	10	450	23	30	53	6	318	0.70	$W_2Q_2^2$
xor5	11	2	22	5	6	11	2	22	0.83	C_4Q_2

two values. It is also shown the ratio of non-terminal and constat nodes $r_n = ntn/cn$. This table shows that, besides depth reduction, we get width reduction, except for 5xp1. Except this function and con1, the area is also reduced. In FNADDs, the ratio of non-terminal and constant nodes r_n is quite more convenient. For example, in xor5, where the size and width remain the same, this ratio r_n is 5.5 and 0.83, respectively.

Table 4.8 compares sizes of SBDDs and FNADDs for adders and multipliers. In this example, FNADDs provide the reduction of all three parameters, depth, size, and width and the reduction is considerable compared to SBDDs. The reduction possibilities in these decision diagrams are compared as the percentage of nodes used from the total of nodes in the corresponding decision trees. It may be seen that these values are comparable, thus, the possibility to do reduction in SBDDs and FNADDs is comparable. In FNADDs, the reduction of area is considerable and shows the advantage of them.

Table 4.9 compares sizes of SBDDs, MTBDDs and FNADDs for adders and multipliers. In SBDDs and MTBDDs, the domain group is C_2^n. The domain group for FNADDs is shown in the table, since depends on n. We can observe that, besides depth reduction, FNADDs permit reduction of width and size. At the same time, FNADDs have smaller area and convenient ratio of non-terminal and constant nodes for applications where reduction of the number of non-terminal nodes at the price of constant nodes may be required. When n increases, these properties of FNADDs increase.

Table 4.10 compares the FNADDs to some other decision diagrams based on spectral transforms. The Arithmetic transform DDs (ACDDs), Walsh transform DDs (WDDs) in $(1, -1)$ coding [27], and Complex-Hadamard Transform DDs [7] are compared with FNADDs for some benchmark functions. ACDDs, WDDs, and CHTDDs are DDs on Abelian groups. Therefore, the depth is equal to n. For rd53 and xor5, the sizes of FNADDs are equal to those of other DDs. However, the number of non-terminal nodes and the width are reduced. In other cases in this example, FNADDs are more efficient with respect to depth, width and size. However, as for other decision diagrams, examples where FNADDs are not efficient certainly can be found. As an ex-

Table 4.8 SBDDs and FNADDs for adders and multipliers.

	adders									
	SBDD				FNADD					
n	s	w	%	a	ntn	cn	s	w	%	a	r_n
2	8	21	22.00	168	4	7	11	2	52.38	22	0.57
3	57	20	42.86	1140	6	7	13	4	31.70	52	0.85
4	103	30	19.84	3090	14	14	28	7	33.73	196	1.00
5	226	62	10.98	14012	18	16	34	7	8.31	238	1.12
6	477	126	5.81	60102	21	12	33	7	4.00	231	1.75

	multipliers									
	SBDD				FNADD					
n	s	w	%	a	ntn	cn	s	w	%	a	r_n
2	19	5	14.28	95	4	10	14	2	9.52	28	0.40
3	63	15	11.27	945	9	20	29	4	9.75	116	0.45
4	159	39	7.51	6201	24	42	66	11	13.25	726	0.57
5	473	114	5.54	53912	37	50	87	17	4.51	1479	0.74
6	788	192	2.34	151296	45	49	94	22	2.68	2068	0.92

Table 4.9 SBDDs, MTBDDs and FNADDs for adders and multipliers.

n	DD	ntn	cn	s	w	a	r_n
				adders			
3	SBDD	55	2	57	20	1140	27.5
	C_2^6	41	15	56	14	784	2.73
	Q_2^2	6	7	13	4	52	0.86
4	SBDD	101	2	103	30	3090	50.50
	C_2^8	113	31	144	30	4320	3.64
	$C_4Q_2^2$	14	15	29	7	203	0.93
5	SBDD	224	2	226	62	14012	112.00
	C_2^{10}	289	63	352	62	21824	4.59
	$C_2Q_2^3$	18	16	34	7	238	1.12
6	SBDD	475	2	477	126	60102	228.50
	C_2^{12}	705	127	832	126	104832	5.55
	Q_2^4	21	12	33	7	231	1.75
				multipliers			
n	DD	ntn	cn	s	w	a	r_n
3	SBDD	61	2	63	15	945	30.5
	C_2^6	56	26	82	28	2296	2.15
	Q_2^2	9	20	29	4	116	0.45
4	SBDD	157	2	159	39	6201	78.50
	C_2^8	240	90	330	120	39600	2.67
	$C_4Q_2^2$	24	46	70	11	770	0.52
5	SBDD	471	2	473	114	53922	235.50
	C_2^{10}	992	340	1332	496	660672	2.92
	$C_2Q_2^3$	46	58	104	23	2392	0.79
6	SBDD	786	2	788	192	151296	393.00
	C_2^{12}	4032	1238	5270	2016	10624320	3.257
	Q_2^4	45	49	94	22	2068	0.92

Table 4.10 Sizes of ACDDs, WDDs, CHTDDs and FNADDs.

f	ACDD						CHTDDs					
	ntn	cn	s	w	a	r_n	ntn	cn	s	w	a	r_n
5xpl	38	11	49	10	490	3.45	127	128	255	64	16320	0.99
bw	31	32	63	16	1008	0.97	31	32	63	16	1008	0.96
con1	40	5	45	10	450	8	115	42	157	56	8792	2.74
rd53	15	6	21	5	105	2.5	15	6	21	5	105	2.5
rd73	27	6	33	6	188	4.5	27	8	36	7	252	3.37
xor5	15	6	21	5	105	2.5	9	2	11	2	22	4.5
f	WDD(1,-1)						FNADD					
	ntn	cn	s	w	a	r_n	ntn	cn	s	w	a	r_n
5xpl	37	13	50	9	450	2.85	39	128	167	18	3006	0.30
bw	31	32	63	16	1008	0.97	9	25	34	4	136	0.36
con1	48	11	59	13	767	4.36	13	12	25	6	150	1.08
rd53	15	5	20	5	100	3.00	7	14	21	3	63	0.50
rd73	28	5	33	7	231	5.6	23	30	53	6	318	0.77
xor5	5	2	7	1	7	2.50	5	6	11	2	22	0.83

Table 4.11 BDDs and FNADDs for Achilles' heel functions.

$n = 2r$	BDD		FNADD		
	worst	best	worst	best	decomposition
4	8	6	13	12	C_2Q_2
6	12	8	37	25	Q_2^2
8	14	10	71	40	$C_4Q_2^2$

ample consider the Achille's heel function $f = x_1y_1 \vee x_2y_2 \vee \cdots \vee x_{2r}y_{2r}$. Table 4.11 shows the sizes of FNADDs for Achille's heel function for different values of n and for two different orderings $x_1, x_2, \ldots, x_{2r}, y_1, y_2, \ldots, y_{2r}$ and $x_1, y_1, x_2, y_2, \ldots, x_ny_n$. This example shows that, as in any other decision diagrams, the sizes of FNADDs greatly depends on the ordering of variables. In this example, the depth reduction in FNADDs is achieved at the price of the increase the size of the FNADDs.

4.9 FOURIER DDS WITH PREPROCESSING

Fourier decision diagrams on non-Abelian groups take advantages from matrix representations of the Fourier transform. In this section, we propose a method which permits to further exploit these properties in decision diagram representations of discrete functions.

We first convert a given number-valued function f into a matrix-valued function f_m by splitting the vector $\mathbf{F}$ of function values for f into subvectors which are arranged as rows of some matrices representing elements of a matrix-valued vector $\mathbf{F}_m$ defining f_m. We represent f_m by a matrix-valued FNADD. Then, in the thus derived FNADD, we represent each matrix-valued constant node by a Shannon decision diagram. In this way, we obtain a number-valued decision diagram representing f in a very compact form.

In what follows, we define the matrix-valued functions and the related Fourier transform. Then, we define the Fourier decision diagrams for matrix-valued functions. This transform is used to define the Fourier decision diagrams for representation of matrix-valued functions. We explain by examples how these decision diagrams can be used to represent number-valued functions.

4.9.1 Matrix-valued functions

A discrete function $f \in P(G)$ is conveniently represented as a vector of its values at all the points of G, thus, f is given as $\mathbf{F} = [f(0), \dots, f(g-1)]^T$.

Definition 4.6 *(Matrix-valued functions)*
A function f defined on G taking values in a set of $(a \times b)$ matrices $M_{a,b}$ over a field P is called a matrix-valued function on G.

The space of all matrix-valued function on G over P is denoted by $P_{a,b}(G)$. A matrix-valued function $f \in P_{a,b}(G)$ is conveniently represented by a vector $\mathbf{F} = [\mathbf{f}(0), \dots, \mathbf{f}(g-1)]^T$, where $\mathbf{f}(i)$ are $(a \times b)$ matrices over P. In this section mainly functions whose values are square matrices are considered, thus, $a = b = r_w$ where r_w is a given number [11]. This dimension will be given the value r_w, where r_w is the dimension of the w-th unitary irreducible representation of G over P [11].

In this vector notation, a number-valued discrete function can be transferred into its matrix-valued equivalent by using the following obvious algorithms.

Algorithm 4.1 *(Vector to matrix transformation (by columns))*

1. *Given a vector $\mathbf{F}$ of order $g = nr_w^2$. Split $\mathbf{F}$ into subvectors consisting of r_w successive elements $f(i)$ of $\mathbf{F}$.*

2. *Write as columns of a matrix the first r_w successive subvectors. Continue with the following r_w subvectors until n matrices of order $(r_w \times r_w)$ are generated.*

3. *The vector $[\mathbf{F}_m]$ of order $g_n = n$ whose elements are the thus generated matrices $\mathbf{f}(i)$ represents the matrix-valued function f_m corresponding to f.*

Algorithm 4.2 *(Vector to matrix transformation (by rows))*
Algorithm 4.1, but with subvectors in 2 written as rows of the generated matrices.

Conversely, an $(r_w \times r_w)$ matrix can be considered as a function on a group of order r_w^2 and, thus, represented by a vector generated by using the following algorithms.

Algorithm 4.3 *(Matrix to vector transformation (by columns))*

1. *Given an $(r_w \times r_w)$ matrix $\mathbf{M}$ over P.*

2. *Generate a vector $\mathbf{V}$ of order r_w^2 by concatenating the columns of $\mathbf{M}$.*

Algorithm 4.4 *(Matrix to vector transformation (by rows))*
Algorithm 4.3, but concatenating the rows of $\mathbf{M}$ in 2.

4.9.2 Fourier transform for matrix-valued functions

It is assumed that the conditions for the existence of the Fourier transform on G over P are satisfied in $P_{r_w,r_w}(G)$.

Denote by $\mathbf{R}_w(x)$ the value of the w-th unitary irreducible representation R_w of G over P at the point $x \in G$. Thus, $\mathbf{R}_w(x)$ stands for an r_w by r_w matrix with elements $R_w^{(i,j)}(x) \in P$, and $i, j \in \{0, \dots, r_w - 1\}$.

The set of all unitary irreducible representations of G forms the dual object Γ of G. In the case of Abelian groups, all the unitary irreducible representations are one-dimensional and, thus, reduce to group characters. In that case Γ expresses the structure of a multiplicative group isomorphic with G. If G is decomposable in the form (2.4), then the cardinality K of Γ can be written as $K = \prod_{i=1}^n K_i$, where K_i is the cardinality of the dual object Γ_i of G_i.

The orthogonality relation for the components of the matrix functions of the dual object Γ is

$$\frac{1}{g} \sum_{x \in G} r_w^{1/2} \overline{R_w^{(i,k)}(x)} r_v^{1/2} R_v^{(j,r)}(x) = \delta_{w,v} \delta_{i,j} \delta_{k,r},$$

where $w, v \in \Gamma$, $1 \le i, k \le r_w$, $1 \le j, r \le r_v$. The bar denotes complex conjugation and δ is the Kronecker delta.

The character χ_w of $\mathbf{R}_w$ is

$$\chi_w(x) = Tr(\mathbf{R}_w(x)).$$

The orthogonality relations for characters are

$$\frac{1}{g} \sum_{x \in G} \overline{\chi_w(x)} \chi_v(x) = \delta_{w,v},$$

$$\frac{1}{g} \sum_{w \in \Gamma} \overline{\chi_w(x)} \chi_w(\zeta) = \frac{\delta_{x,\zeta}}{p_x},$$

where p_x is the number of elements in the conjugate class of G that contains x. Recall that the number of conjugate classes of a group G is equal to the number of elements in the set Γ.

Definition 4.7 *The direct and inverse Fourier transform of $f \in P_{k,m}(G)$ are defined respectively by*

$$\mathbf{S}_f(w) = (S_f^{(i,j)}(w)) = r_w g^{-1} \sum_{u=0}^{g-1} f^{(i,j)}(u)\mathbf{R}_w(u^{-1}), \tag{4.1}$$

$$\mathbf{f}(x) = (f^{(i,j)}(x)) = \sum_{w=0}^{K-1} Tr(\mathbf{S}_f^{(i,j)}(w)\mathbf{R}_w(x)), \tag{4.2}$$

where K is the number of unitary irreducible representations $\mathbf{R}_w$ of G over P.

Notice that for a function $f \in P_{k,m}(G)$ with group representations of order r_w, the Fourier coefficients are in M_{kr_w,mr_w}.

Here and in the sequel we shall assume, without explicitly saying so, that all arithmetical operations are carried out in the field P.

In the matrix notation, the Fourier transform matrix on finite groups is given as follows [11].

Definition 4.8 *The Fourier transform on G is defined by the transform matrix $[\mathbf{R}]^{-1}$ inverse to the matrix of basis functions defined as group representations for G, i.e., $[\mathbf{R}] = \frac{r_w}{g}[\mathbf{R}_0, \mathbf{R}_1, \cdots \mathbf{R}_{K-1}]$, where*

$$\mathbf{R}_0 = \begin{bmatrix} R_0^{(0,0)}(0) & \cdots & R_0^{(0,r_w-1)}(0), \\ \vdots & \vdots & \vdots \\ R_0^{(r_w-1,0)}(g-1) & \cdots & R_0^{(r_w-1,r_w-1)}(g-1) \end{bmatrix},$$

$$\mathbf{R}_1 = \begin{bmatrix} R_1^{(0,0)}(0) & \cdots & R_1^{(0,r_w-1)}(0), \\ \vdots & \vdots & \vdots \\ R_1^{(r_w-1,0)}(g-1) & \cdots & R_1^{(r_w-1,r_w-1)}(g-1) \end{bmatrix},$$

$$\cdots$$

$$\mathbf{R}_{K-1} = \begin{bmatrix} R_{K-1}^{(0,0)}(0) & \cdots & R_{K-1}^{(0,r_w-1)}(0), \\ \vdots & \vdots & \vdots \\ R_{K-1}^{(r_w-1,0)}(g-1) & \cdots & R_{K-1}^{(r_w-1,r_w-1)}(g-1) \end{bmatrix},$$

where $[\mathbf{R}] \in M_{g,r_w^2}$ and $R_w^{(i,j)}(\cdot)$ denotes the (i,j)-th element of $\mathbf{R}_w(\cdot)$, $1 \leq p,q \leq r_w$. Thus, $\mathbf{R} \in M_{g,g}$.

Example 4.11 *The Fourier transform on S_3 over $GF(11)$ introduced in Example 2.2 is given by*

$$[\mathbf{R}]^{-1} = \frac{1}{6} \begin{bmatrix} 1 & 1 & 1 & 1 & 1 & 1 \\ 1 & 1 & 1 & -1 & -1 & -1 \\ 2\mathbf{I} & 2\mathbf{B} & 2\mathbf{A} & 2\mathbf{C} & 2\mathbf{D} & 2\mathbf{E} \end{bmatrix},$$

with notations as in Table 2.2 and Table 2.5, respectively. Notice that in the case of representations over $GF(11)$, the normalization factor is $1/6 = 2$.

Example 4.12 *The Fourier transform for functions on Q_2 over C is defined by group representations in Table 2.9 and it was introduced in Example 2.3. For convenience, it is repeated here, thus,*

$$[\mathbf{Q_2}]^{-1} = \frac{1}{8} \begin{bmatrix} 1 & 1 & 1 & 1 & 1 & 1 & 1 & 1 \\ 1 & -1 & 1 & -1 & 1 & -1 & 1 & -1 \\ 1 & 1 & 1 & 1 & -1 & -1 & -1 & -1 \\ 1 & -1 & 1 & -1 & -1 & 1 & -1 & 1 \\ 2\mathbf{I} & 2i\mathbf{B} & -2\mathbf{I} & 2i\mathbf{A} & 2\mathbf{E} & 2i\mathbf{D} & 2\mathbf{C} & -2i\mathbf{D} \end{bmatrix},$$

with notations as in Table 2.9.

Example 4.13 *Function xor5 is a five variable function defined as $xor5 = x_1 \oplus x_2 \oplus x_3 \oplus x_4 \oplus x_5$ and is given by the truth-vector*

$$\mathbf{F} = [0,1,1,0,1,0,0,1,1,0,0,1,0,1,1,0,1,0,0,1,0,1,1,0,0,1,1,0,1,0,0,1]^T.$$

By using the algorithm 4.2, xor5 transfers into the matrix-valued function f_m given by the vector

$$[\mathbf{F}_m] = \begin{bmatrix} \begin{bmatrix} 0 & 1 \\ 1 & 0 \end{bmatrix}, \begin{bmatrix} 1 & 0 \\ 0 & 1 \end{bmatrix}, \begin{bmatrix} 1 & 0 \\ 0 & 1 \end{bmatrix}, \begin{bmatrix} 0 & 1 \\ 1 & 0 \end{bmatrix}, \\ \begin{bmatrix} 1 & 0 \\ 0 & 1 \end{bmatrix}, \begin{bmatrix} 0 & 1 \\ 1 & 0 \end{bmatrix}, \begin{bmatrix} 0 & 1 \\ 1 & 0 \end{bmatrix}, \begin{bmatrix} 1 & 0 \\ 0 & 1 \end{bmatrix} \end{bmatrix}^T.$$

The Fourier coefficients are calculated as

$$\begin{aligned} \mathbf{S}_f(w) = (S_f^{(i,j)}(w)) &= \frac{2}{8}(f^{(i,j)}(0)\mathbf{R}_w(0) + f^{(i,j)}(1)\mathbf{R}_w(3) + f^{(i,j)}(2)\mathbf{R}_w(2) \\ &+ f^{(i,j)}(3)\mathbf{R}_w(1) + f^{(i,j)}(4)\mathbf{R}_w(6) + f^{(i,j)}(5)\mathbf{R}_w(7) \\ &+ f^{(i,j)}(6)\mathbf{R}_w(4) + f^{(i,j)}(7)\mathbf{R}_w(5). \end{aligned}$$

The Fourier spectrum of f_m is given by

$$[\mathbf{S}_f] = \begin{bmatrix} \mathbf{S}_f(0) & \mathbf{S}_f(1) & \mathbf{S}_f(2) & \mathbf{S}_f(3) & \mathbf{S}_f(4) \end{bmatrix}^T,$$

where

$$\mathbf{S}_f(0) = \frac{1}{8} \begin{bmatrix} 4 & 4 \\ 4 & 4 \end{bmatrix},$$

$$\mathbf{S}_f(1) = \mathbf{S}_f(2) = \mathbf{S}_f(3) = \begin{bmatrix} 0 & 0 \\ 0 & 0 \end{bmatrix},$$

$$\mathbf{S}_f(4) = \frac{1}{4} \begin{bmatrix} \begin{bmatrix} -1-i & 1-i \\ -1-i & -1+i \end{bmatrix} & \begin{bmatrix} 1+i & -1+i \\ 1+i & 1-i \end{bmatrix} \\ \begin{bmatrix} 1+i & -1+i \\ 1+i & 1-i \end{bmatrix} & \begin{bmatrix} -1-i & 1-i \\ -1-i & -1+i \end{bmatrix} \end{bmatrix},$$

or in matrix notation

$$[\mathbf{S}_f] = \frac{1}{8} \begin{bmatrix} \begin{bmatrix} 4 & 4 \\ 4 & 4 \end{bmatrix} \\ \begin{bmatrix} 0 & 0 \\ 0 & 0 \end{bmatrix} \\ \begin{bmatrix} 0 & 0 \\ 0 & 0 \end{bmatrix} \\ \begin{bmatrix} 0 & 0 \\ 0 & 0 \end{bmatrix} \\ 2 \begin{bmatrix} \begin{bmatrix} -1-i & 1-i \\ -1-i & -1+i \end{bmatrix} & \begin{bmatrix} 1+i & -1+i \\ 1+i & 1-i \end{bmatrix} \\ \begin{bmatrix} 1+i & -1+i \\ 1+i & 1-i \end{bmatrix} & \begin{bmatrix} -1-i & 1-i \\ -1-i & -1+i \end{bmatrix} \end{bmatrix} \end{bmatrix}.$$

The function can be reconstructed by the application of the relation 4.2 as follows:

$$f(\zeta) = (f^{(i,j)}(\zeta)) = \sum_{w=0}^{K-1} Tr(\mathbf{S}_f^{(i,j)}(w)\mathbf{R}_w(\zeta)).$$

Thus,

$$\begin{aligned} f^{(i,j)}(\zeta) &= Tr(\mathbf{S}_f^{(i,j)}(0)\mathbf{R}_0(\zeta)) + Tr(\mathbf{S}_f^{(i,j)}(1)\mathbf{R}_1(\zeta)) \\ &\quad Tr(\mathbf{S}_f^{(i,j)}(2)\mathbf{R}_2(\zeta)) + Tr(\mathbf{S}_f^{(i,j)}(3)\mathbf{R}_3(\zeta)) + Tr(\mathbf{S}_f^{(i,j)}\mathbf{R}_4(\zeta)). \end{aligned}$$

We will illustrate this procedure by calculating the value of the function $\mathbf{f}(0)$ *and* $\mathbf{f}(7)$.

$$
\begin{aligned}
8f^{(0,0)}(0) &= Tr(4 \cdot 1) + Tr(0 \cdot 1) + Tr(0 \cdot 1) + Tr(0 \cdot 1) \\
&\quad + Tr\left(2 \begin{bmatrix} -1-i & 1-i \\ -1-i & -1+i \end{bmatrix} \begin{bmatrix} 1 & 0 \\ 0 & 1 \end{bmatrix}\right) \\
&= 4 + 0 + 0 + 0 - 4 = 0, \\
8f^{(0,1)}(0) &= Tr(4 \cdot 1) + Tr(0 \cdot 1) + Tr(0 \cdot 1) + Tr(0 \cdot 1) \\
&\quad + Tr\left(2 \begin{bmatrix} 1+i & -1+i \\ 1+i & 1-i \end{bmatrix} \begin{bmatrix} 1 & 0 \\ 0 & 1 \end{bmatrix}\right) \\
&= 4 + 0 + 0 + 0 + 4 = 8, \\
8f^{(1,0)}(0) &= Tr(4 \cdot 1) + Tr(0 \cdot 1) + Tr(0 \cdot 1) + Tr(0 \cdot 1) \\
&\quad + Tr\left(2 \begin{bmatrix} 1+i & -1+i \\ 1+i & 1-i \end{bmatrix} \begin{bmatrix} 1 & 0 \\ 0 & 1 \end{bmatrix}\right) \\
&= 4 + 0 + 0 + 0 + 4 = 8, \\
8f^{(1,1)}(0) &= Tr(4 \cdot 1) + Tr(0 \cdot 1) + Tr(0 \cdot 1) + Tr(0 \cdot 1) \\
&\quad + Tr\left(2 \begin{bmatrix} -1-i & 1-i \\ -1-i & -1+i \end{bmatrix} \begin{bmatrix} 1 & 0 \\ 0 & 1 \end{bmatrix}\right) \\
&= 4 + 0 + 0 + 0 - 4 = 0.
\end{aligned}
$$

$$
\begin{aligned}
8f^{(0,0)}(7) &= Tr(4 \cdot 1) + Tr(0 \cdot (-1)) + Tr(0 \cdot (-1)) + Tr(0 \cdot 1) \\
&\quad + Tr\left(2 \begin{bmatrix} -1-i & 1-i \\ -1-i & -1+i \end{bmatrix} \begin{bmatrix} 0 & i \\ i & 0 \end{bmatrix}\right) \\
&= 4 + 0 + 0 + 0 + 4 = 8, \\
8f^{(0,1)}(7) &= Tr(4 \cdot (-1)) + Tr(0 \cdot (-1)) + Tr(0 \cdot 1) + Tr(0 \cdot 1) \\
&\quad + Tr\left(2 \begin{bmatrix} 1+i & -1+i \\ 1+i & 1-i \end{bmatrix} \begin{bmatrix} 0 & i \\ i & 0 \end{bmatrix}\right) \\
&= 4 + 0 + 0 + 0 - 4 = 0, \\
8f^{(1,0)}(7) &= Tr(4 \cdot 1) + Tr(0 \cdot (-1)) + Tr(0 \cdot (-1)) + Tr(0 \cdot 1) \\
&\quad + Tr\left(2 \begin{bmatrix} 1+i & -1+i \\ 1+i & 1-i \end{bmatrix} \begin{bmatrix} 0 & i \\ i & 0 \end{bmatrix}\right) \\
&= 4 + 0 + 0 + 0 - 4 = 0, \\
8f^{(1,1)}(7) &= Tr(4 \cdot 1) + Tr(0 \cdot (-1)) + Tr(0 \cdot (-1)) + Tr(0 \cdot 1) \\
&\quad + Tr\left(2 \begin{bmatrix} -1-i & 1-i \\ -1-i & -1+i \end{bmatrix} \begin{bmatrix} 0 & i \\ i & 0 \end{bmatrix}\right) \\
&= 4 + 0 + 0 + 0 + 4 = 8.
\end{aligned}
$$

The other function values are determined in the same way.

4.10   FOURIER DECISION TREES WITH PREPROCESSING

Decision trees are defined by using some function expansions [16]. For example, the Shannon tree is defined by using the Shannon decomposition rule [3]. *Walsh transform decision trees* (WDTs) are defined by using the function expansion derived from the Walsh transform [27]. The discrete Walsh transform is the Fourier transform on the dyadic groups, since the Walsh functions are the group characters of the dyadic groups. Thus, WDDs are an example of Fourier DDs on these particular Abelian groups. The concept was extended to any finite not necessarily Abelian group through Fourier DDs [22] that were obtained by using the function decomposition determined by the Fourier transform on finite groups to define the Fourier transform decision trees. In this section we extend the approach to matrix-valued functions.

The Fourier transform on a finite group G decomposable in the form (2.4) may be considered as the n-dimensional Fourier transform on the constituent subgroups G_i. The Fourier transforms on G_i will be used to define a mapping performed at the nodes of a Fourier decision tree to associate a function to the DT.

Definition 4.9 *Denote by* $\mathbf{S}_f$ *the Fourier transform coefficients on a finite not necessarily Abelian group G representable in the form (2.4). The Fourier decision tree on G over P is defined as the decision tree in which the nodes at the i-th level represent functions*

$$\mathbf{f}(x_i) = (f^{(r,s)}(x_i)) = \sum_{w_i \in \Gamma_i} Tr(\mathbf{S}_f^{(r,s)}(w_i)\mathbf{R}_{w_i}(x_i)), \quad x_i \in G_i.$$

This definition differs from Definition 4.1 in the property that all the Fourier coefficients are matrix-valued.

With this definition, Fourier decision trees are the decision trees on G in which each path from the root node up to the constant nodes corresponds to a unitary irreducible representation of G. A node at the i-th level has K_i outgoing edges denoted by $\mathbf{R}_{w_i}(j), j \in \Gamma_i$. In the figures of decision trees we use the short notation w_i^j for $\mathbf{R}_{w_i}(j)$ or simply j when the context does not allow misunderstandings.

For a given f, the values of constant nodes in the Fourier decision diagrams are the Fourier coefficients of f. It follows that, as in decision diagrams on Abelian groups, we perform the inverse Fourier transform in the determination of f represented by a given Fourier decision tree by following the labels at the edges in the DT [21].

In application of the thus defined Fourier decision trees to number-valued functions, these functions should be first transferred into their matrix-valued equivalents. Therefore, such decision trees will be called *the Fourier decision trees on non-Abelian groups with preprocessing* (FNAPDTs).

The structure of matrix-valued FNAPDTs (mvFNAPDTs) is determined as follows. This statement points out the difference with decision trees on Abelian groups.

Observation 4.6 *The number of levels in a FNAPDTs is equal to the number of constituent subgroups G_i in the domain group G of f. The number of outgoing edges of nodes at the i-th level is determined by the cardinality K_i of the dual object Γ_i of* *the corresponding subgroup G_i. The number of constant nodes in FNAPDTs is equal to the cardinality of the dual object Γ of G.*

The difference with the FNADTs is that in FNAPDTs all the constant nodes are matrix-valued. In FNADTs most of the constant nodes are number-valued. Those corresponding to group representations of orders $r_w > 1$ are the matrix-valued. That difference originates form the difference in definition of the Fourier transform for number-valued and matrix-valued functions on non-Abelian groups.

The matrix-valued coefficients in FNADTs and FNAPDTs may be also represented by decision trees by using the method of representation of matrices by the decision diagrams [5]. In that way, we derive the integer-valued or complex-valued Fourier decision trees on finite non-Abelian groups, depending on the field P over which the group representations are taken. In that case, the number of constant nodes in a Fourier DT on a non-Abelian group is equal to g_i, that is, it is the same as in any DT on Abelian group of the same order. Comparing to the matrix-valued Fourier decision trees, the number of levels in an integer-valued or complex-valued Fourier DT is increased to represent the matrix-valued constant nodes. However, comparing to decision trees on Abelian groups, non-terminal nodes still may be saved, since always $K_i < g_i$.

The non-terminal nodes in the Fourier decision trees are denoted by F_i for Abelian groups and on non-Abelian groups by FA_i and FNA_i for decision trees with pre-processing. The index i denotes the number of outgoing edges and is equal to the cardinality K_i of the dual object Γ_i of G_i.

4.11 FOURIER DECISION DIAGRAMS WITH PREPROCESSING

Decision diagrams are defined by the reduction of the decision trees. The reduction rules are chosen depending on the range of the functions represented. Therefore, *the Fourier decision diagrams with preprocessing* (FNAPDDs) are defined as follows.

Definition 4.10 *Fourier decision diagrams with preprocessing are the decision diagrams that are derived from the Fourier decision trees with preprocessing by using the generalized BDD reduction rules. A FNAPDDs is called reduced if further reduction with the same rules is impossible.*

By definition, Fourier decision trees are graphical representations of the Fourier expansion of a given function f. Therefore, they are canonic representations since the Fourier expansions are canonic representations. The same applies to Fourier decision diagrams, since the generalized BDD reduction rules do not destroy or diminish the information content in the decision trees.

In decision diagram representations of integer-valued functions nodes with negative edges are usually used to reduce the complexity of a decision diagram. A negative edge indicates that the value or subfunction v represented by the node where the edge points to, should be multiplied by -1. Thus, two nodes with values v and $-v$ and

Table 4.12   Labels of edges.

v	l
$y + iz$	j
$y - iz$	j^*
$-y - iz$	$-j$
$-y + iz$	$-j^*$
iz	ij

the incoming edges j and k, may be replaced by one node whose value is v and two incoming edges are labeled by j and $-k$.

In various decision diagrams representing complex-valued functions on finite groups, as well as in Fourier decision diagrams over C for whatever integer or complex-valued function, there are nodes whose outgoing edges point to nodes representing pure imaginary values or nodes representing mutually complex-conjugate values. In these decision diagrams, by the same justification as for the nodes with negative edges, we can consider imaginary edges denoting the multiplication with the imaginary unity i. Thus, two nodes whose values are v and iv with the incoming edges labeled by j and k, respectively, can be represented by one node whose value is v and the two incoming edges are labeled by j and ik.

In a similar way, simplification of decision diagrams representations may be done by introduction of the node with complex-conjugate edges permitting to reduce nodes representing a value or subfunction v and its complex-conjugate v^*. Labels of corresponding imaginary and complex-conjugate edges are shown in Table 4.12. In this table, v denotes the value or subfunction represented by a node and l is the label at the edge corresponding to $\mathbf{R}_{w_i}(j)$ for non-Abelian and $x_i = j$ for Abelian groups. The last row in the table shows, that if there is a node showing the value z and another showing the imaginary value iz, which is pointed by an edge j, we can redirect j to the node z and change the label into $i \cdot j$, where $i = \sqrt{-1}$.

4.12   CONSTRUCTION OF FNAPDD

FNADDs take advantage from the matrix notation and thus provide compact representation of f in the spectral domain. In FNAPDDs the same advantages are extended to both original and spectral domain.

If a given function f on a group G of order g is transferred into a matrix-valued function on a finite non-Abelian group G_n of order $g_n = g/r_w$, where r_w is the greatest order of the group representations of G_n, then, f is represented by the FNAPDD on G_n. The elements of the matrix-valued constant nodes in this FNAPDDs may be considered as functions on an Abelian group of order r_w and represented by the

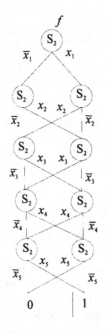

Fig. 4.41 BDD representation for xor5.

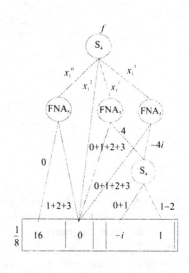

Fig. 4.42 FNADD for xor5.

decision diagrams on this group. In that way a decision diagram with small depth and width may be produced for f.

Construction of FNAPDDs is explained and these decision diagrams compared with FNADDs by the following examples.

Example 4.14 *Function* xor5 *is given by the truth-vector*

$$\mathbf{F} = [0, 1, 1, 0, 1, 0, 0, 1, 1, 0, 0, 1, 0, 1, 1, 0, 1, 0, 0, 1, 0, 1, 1, 0, 0, 1, 1, 0, 1, 0, 0, 1]^T,$$

and can be represented by the BDD in Figure 4.41. xor5 *can be considered as a function on the group* $G_{32} = C_4 \times Q_2$. *The group representations of* C_4 *over the complex field are given by the complex Walsh matrix*

$$\mathbf{CW} = \begin{bmatrix} 1 & 1 & 1 & 1 \\ 1 & -i & -1 & i \\ 1 & -1 & 1 & -1 \\ 1 & i & -1 & -i \end{bmatrix}$$

Since, $\mathbf{CW}^{-1} = \frac{1}{4}\mathbf{CW}$, *the Fourier transform on* $G_{32} = C_4 \times Q_2$ *is defined by the matrix*

$$\mathbf{R}_{32}^{-1} = \mathbf{CW}^{-1} \otimes \mathbf{Q}_2^{-1}.$$

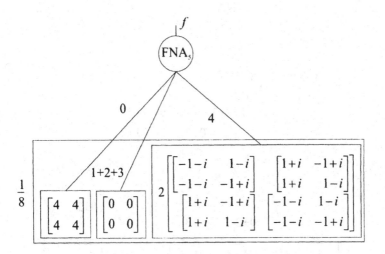

Fig. 4.43 Matrix-valued FNAPDD for xor5.

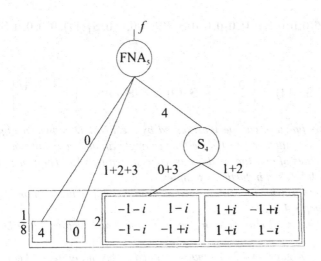

Fig. 4.44 FNAPDD for xor5 with elements of mv nodes.

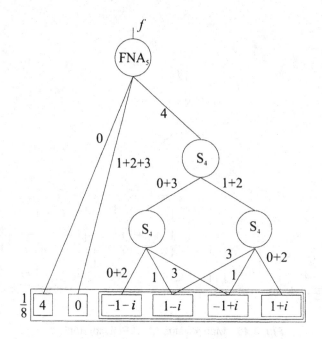

Fig. 4.45 Complex-valued FNAPDD for xor5.

The Fourier spectrum of xor5 *on this group is given by*

$$[\mathbf{S}_f] = \frac{1}{32}[16, 0, 0, 0, \mathbf{S}_f(4), 0, 0, 0, 0, \mathbf{S}_f(9), 0, 0, 0, 0, \mathbf{S}_f(14), 0, 0, 0, 0, \mathbf{S}_f(19)]^T,$$

where

$$\mathbf{S}_f(4) = \mathbf{S}_f(14) = \begin{bmatrix} 0 & 0 \\ 0 & 0 \end{bmatrix}, \mathbf{S}_f(9) = -i\mathbf{S}_f(19) = \begin{bmatrix} -i & 1 \\ -i & -1 \end{bmatrix}.$$

Therefore, this function can be represented by the FNADD shown in Figure 4.42. In this figure, the number 1/8 before the rectangle around constant nodes denotes that all the values of constant nodes has to be scaled by 1/8. The same notational convention will be used in further examples.

The Fourier spectrum for xor5 *viewed as a matrix-valued function on Q_2 is determined in Example 4.13.*

It follows that xor5 *may be represented by the matrix-valued FNAPDD in Figure 4.43.*

By using the Algorithm 4.2, the matrix-valued nodes in Figure 4.43 can be represented by Multi-terminal decision diagrams (MTDDs) on C_4^2 or C_2^4. From (4.2), to read the function values for f, we should work with matrix elements of the matrix-valued Fourier spectral coefficients, i.e., with submatrices $\mathbf{S}_w^{(i,j)}(w)$ of order r_w. It follows, that in a mvFNAPDD, it is sufficient to represent the matrix-valued coeffi-

Table 4.13 Complexity of representation of xor5 in terms of the number of levels, non-terminal nodes (ntn), constant nodes (cn), and sizes.

xor5	d	ntn	cn	s
BDD	6	9	2	11
Matrix-valued FNADD	3	2	3	5
Complex-valued FNADD	4	4	4	8
Matrix-valued FNAPDD	2	1	3	4
Complex-valued FNAPDD	4	4	6	10

cients until their submatrix elements of order r_w. However, these elements can be further represented by complex-valued decision diagrams.

Therefore, xor5 may be represented by the complex-valued FNAPDD in Figure 4.45.

Complexity of representation of xor5 by BDD FNADD and FNAPDD is compared in Table 4.13. This table shows the depth (d), width (w), number of non-terminal nodes (ntn), constant nodes (cn), and the size (s) for different decision diagrams for xor5.

Example 4.14 shows that compared to BDD and FNADD for xor5, FNAPDD reduces the depth, width and size of the DD. The number of non-terminal nodes is also reduced. The price is the increased number of constant nodes. However, that saving of non-terminal nodes (7 instead 9 for BDD) is in some applications more important than the increase of the number of constant nodes (4 instead of 2).

The following example illustrates the use of FNAPDDs when the domain group G is represented as the product of Abelian and non-Abelian groups.

Example 4.15 *Figure 4.46 shows BDD for the switching function $f(x_1, x_2, x_3, x_4, x_5)$ given by the following truth-table (taken from [14]).*

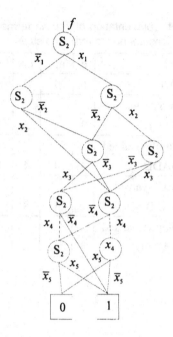

Fig. 4.46 BDD for f in Example 4.15.

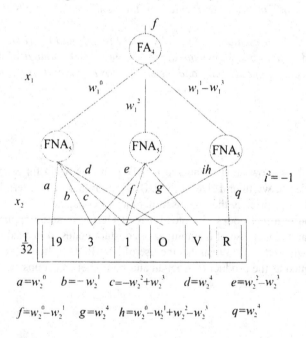

$a=w_2^0 \quad b=-w_2^1 \quad c=-w_2^2+w_2^3 \quad d=w_2^4 \quad e=w_2^2-w_2^3$

$f=w_2^0-w_2^1 \quad g=w_2^4 \quad h=w_2^0-w_2^1+w_2^2-w_2^3 \quad q=w_2^4$

Fig. 4.47 Matrix-valued FNADD for f in Example 4.15.

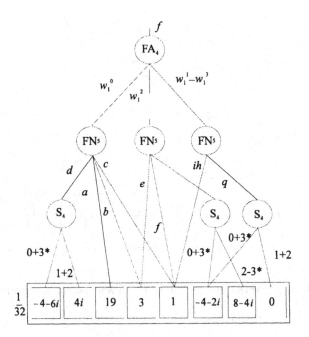

Fig. 4.48 Complex-valued FNADD for f in Example 4.15.

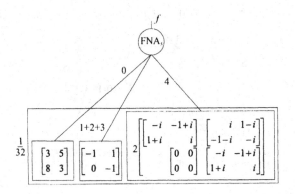

Fig. 4.49 Matrix-valued FNAPDD for f in Example 4.15.

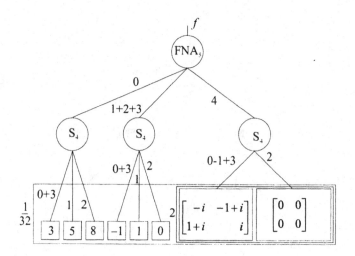

Fig. 4.50 FNAPDD for f in Example 4.15 with elements of mv nodes.

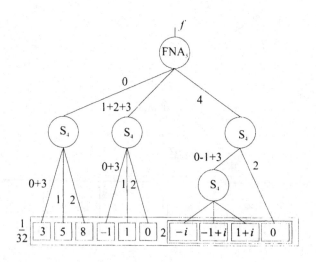

Fig. 4.51 Complex-valued FNAPDD for f in Example 4.15.

				$x_1x_2x_3$					
			0	0	0	0	1	1	1	1
			0	0	1	1	0	0	1	1
			0	1	0	1	0	1	0	1
	0	0	0	1	0	0	0	1	1	0
x_4x_5	0	1	1	1	1	1	1	1	1	1
	1	0	1	0	1	1	1	0	0	1
	1	1	0	1	0	0	0	1	1	0

This function can be alternatively considered as a function on the group $G_{32} = C_4 \times Q_2$. The Fourier spectrum of f is thus given by

$$\mathbf{S}_f = \frac{1}{32}[19, -3, -1, 1, \mathbf{O}, i, -i, i, -i, \mathbf{R}, 1, -1, -3, 3, \mathbf{V}, -i, i, -i, i, -\mathbf{R}]^T,$$

where

$$\mathbf{O} = \begin{bmatrix} -4 - 6i & 4i \\ 4i & -4 + 6i \end{bmatrix}, \mathbf{R} = \begin{bmatrix} -2 + 4i & 0 \\ 0 & 2 + 4i \end{bmatrix},$$
$$\mathbf{V} = \begin{bmatrix} -4 - 2i & 8 - 4i \\ -8 - 4i & -4 + 2i \end{bmatrix}.$$

Therefore, this function can be represented by the matrix-valued FNADD shown in Figure 4.47. In this decision diagram, the matrix-valued nodes are transferred into functions on C_4 by using the algorithm A_2 and represented by the nodes S_4 with four outgoing edges. Figure 4.48 shows the thus derived complex-valued FNADD of f.

This function f can be transferred into the matrix-valued function f_m on Q_2 given by the vector

$$[\mathbf{F}_m] = \left[\begin{bmatrix} 0 & 1 \\ 1 & 0 \end{bmatrix}, \begin{bmatrix} 1 & 0 \\ 1 & 1 \end{bmatrix}, \begin{bmatrix} 0 & 1 \\ 1 & 0 \end{bmatrix}, \begin{bmatrix} 0 & 1 \\ 1 & 0 \end{bmatrix}, \right.$$
$$\left. \begin{bmatrix} 0 & 1 \\ 1 & 0 \end{bmatrix}, \begin{bmatrix} 1 & 0 \\ 1 & 1 \end{bmatrix}, \begin{bmatrix} 1 & 0 \\ 1 & 1 \end{bmatrix}, \begin{bmatrix} 0 & 1 \\ 1 & 0 \end{bmatrix} \right]^T.$$

The Fourier spectrum of f_m is given by

$$[\mathbf{S}_f] = [\mathbf{S}_f(0), \mathbf{S}_f(1), \mathbf{S}_f(2), \mathbf{S}_f(3), \mathbf{S}_f(4)]^T,$$

where

$$\mathbf{S}_f(0) = \frac{1}{8} \begin{bmatrix} 3 & 5 \\ 8 & 3 \end{bmatrix},$$

$$\mathbf{S}_f(1) = \mathbf{S}_f(2) = \mathbf{S}_f(3) = \frac{1}{8} \begin{bmatrix} -1 & 1 \\ 0 & -1 \end{bmatrix},$$

Table 4.14 Complexities of representing f in Example 4.15 in terms of the number of levels, non-terminal nodes (ntn), constant nodes, (cn), and sizes.

Number of	d	ntn	cn	s
BDD	6	9	2	11
Matrix-valued FNADD	3	4	6	10
Complex-valued FNADD	4	7	8	15
Matrix-valued FNAPDD	2	1	5	4
Complex-valued FNAPDD	4	5	9	15

$$\mathbf{S}_f(4) = \frac{1}{4} \left[\begin{array}{cc} \left[\begin{array}{cc} -i & -1+i \\ 1+i & i \end{array} \right] & \left[\begin{array}{cc} i & 1-i \\ -1-i & -i \end{array} \right] \\ \left[\begin{array}{cc} 0 & 0 \\ 0 & 0 \end{array} \right] & \left[\begin{array}{cc} -i & -1+i \\ 1+i & i \end{array} \right] \end{array} \right].$$

It follows that f can be represented by the matrix-valued FNAPDD in Figure 4.49. Figure 4.50 shows FNAPDD with elements of matrix-valued nodes represented by matrix-valued MTDDs. If these elements are represented by decision diagrams, the function f can be represented by the complex-valued FNAPDD in Figure 4.51. Complexity of representation of the considered function by BDD, FNADD and FNAPDD is compared in Table 4.14.

In this example, FNAPDD permits reduction of the depth and the number of non-terminal nodes without exceeding the size of BDD for f. This reduction is obtained as a trade-off between the complexity of the overall decision diagram and the complexities of the nodes of the decision diagram.

The following example illustrates construction of FNAPDDs over finite fields.

Example 4.16 *Figure 4.52 shows MTDD representation of the function f given by the following truth-table (taken from [14]).*

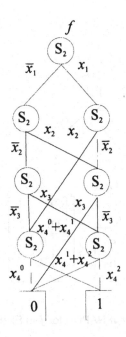

Fig. 4.52 MTDD for f in Example 4.16 on $G_{24} = C_2 \times C_2 \times C_2 \times C_3$.

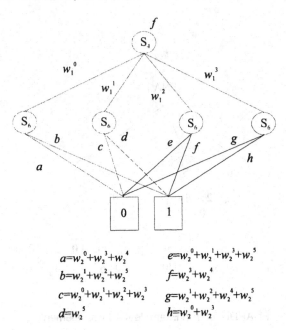

$$a = w_2^0 + w_2^3 + w_2^4 \qquad e = w_2^0 + w_2^1 + w_2^3 + w_2^5$$
$$b = w_2^1 + w_2^2 + w_2^5 \qquad f = w_2^3 + w_2^4$$
$$c = w_2^0 + w_2^1 + w_2^2 + w_2^3 \qquad g = w_2^1 + w_2^2 + w_2^4 + w_2^5$$
$$d = w_2^5 \qquad h = w_2^0 + w_2^3$$

Fig. 4.53 MTDD for f in Example 4.16 on $G_{24} = C_4 \times C_6$.

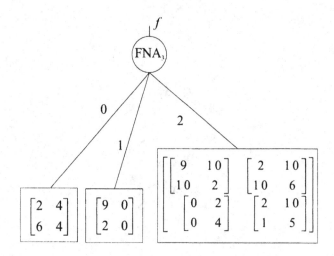

Fig. 4.54 mvFNAPDD for f in Example 4.16.

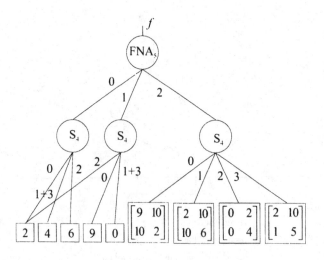

Fig. 4.55 FNAPDD for f in Example 4.16 with elements of mv nodes.

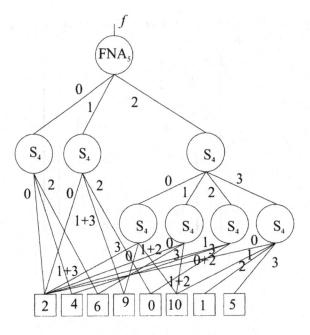

Fig. 4.56 ivFNAPDD for f in Example 4.16.

$x_1x_2x_3x_4$	f	$x_1x_2x_3x_4$	f	$x_1x_2x_3x_4$	f	$x_1x_2x_3x_4$	f
0000	0	0100	0	1000	0	1100	1
0001	1	0101	0	1001	0	1101	0
0002	1	0102	0	1002	0	1102	0
0010	0	0110	0	1010	1	1110	1
0011	0	0111	0	1011	1	1111	0
0012	1	0112	1	1012	0	1112	0

If f is regarded as a function on $G_{24} = C_4 \times C_6$, where C_4 and C_6 are the groups of orders 4 and 6, it can be represented by the MTDD with two-levels in Figure. 4.53.

If f is transferred into the matrix-valued function f_m on the symmetric group of permutations S_3 over $GF(11)$ by using the Algorithm 4.1, it is given by

$$[\mathbf{F}_m] = \left[\begin{bmatrix} 0 & 1 \\ 1 & 0 \end{bmatrix}, \begin{bmatrix} 0 & 0 \\ 1 & 0 \end{bmatrix}, \begin{bmatrix} 0 & 0 \\ 0 & 1 \end{bmatrix}, \begin{bmatrix} 0 & 0 \\ 0 & 1 \end{bmatrix}, \begin{bmatrix} 1 & 1 \\ 0 & 0 \end{bmatrix}, \begin{bmatrix} 0 & 0 \\ 1 & 0 \end{bmatrix}\right]^T.$$

By using the Fourier transform for matrix-valued functions on S_3, the Fourier transform of f_m on S_3 is given by

$$[\mathbf{S}_f] = 2 \begin{bmatrix} \begin{bmatrix} 1 & 2 \\ 3 & 2 \end{bmatrix} \\ \begin{bmatrix} 10 & 0 \\ 1 & 0 \end{bmatrix} \\ 2 \begin{bmatrix} \begin{bmatrix} 5 & 8 \\ 8 & 6 \end{bmatrix} & \begin{bmatrix} 6 & 8 \\ 8 & 7 \end{bmatrix} \\ \begin{bmatrix} 0 & 6 \\ 0 & 1 \end{bmatrix} & \begin{bmatrix} 6 & 8 \\ 3 & 4 \end{bmatrix} \end{bmatrix} \end{bmatrix}$$

$$= \begin{bmatrix} \begin{bmatrix} 2 & 4 \\ 6 & 4 \end{bmatrix} \\ \begin{bmatrix} 9 & 0 \\ 2 & 0 \end{bmatrix} \\ \begin{bmatrix} \begin{bmatrix} 9 & 10 \\ 10 & 2 \end{bmatrix} & \begin{bmatrix} 2 & 10 \\ 10 & 6 \end{bmatrix} \\ \begin{bmatrix} 0 & 2 \\ 0 & 4 \end{bmatrix} & \begin{bmatrix} 2 & 10 \\ 1 & 5 \end{bmatrix} \end{bmatrix} \end{bmatrix} .$$

Thus, the spectrum is

$$[\mathbf{S}_f] = [\mathbf{S}_f(0), \mathbf{S}_f(1), \mathbf{S}_f(2)]^T ,$$

where

$$\mathbf{S}_f(0) = \begin{bmatrix} 2 & 4 \\ 6 & 4 \end{bmatrix} ,$$

$$\mathbf{S}_f(1) = \begin{bmatrix} 9 & 0 \\ 2 & 0 \end{bmatrix} ,$$

$$\mathbf{S}_f(2) = \begin{bmatrix} \begin{bmatrix} 9 & 10 \\ 10 & 2 \end{bmatrix} & \begin{bmatrix} 2 & 10 \\ 10 & 6 \end{bmatrix} \\ \begin{bmatrix} 0 & 2 \\ 0 & 4 \end{bmatrix} & \begin{bmatrix} 2 & 10 \\ 1 & 5 \end{bmatrix} \end{bmatrix} .$$

Therefore, f can be represented by the matrix-valued FNAPDD in Figure 4.54. The same FNAPDD is shown in Figure 4.55 with elements of the matrix-valued nodes

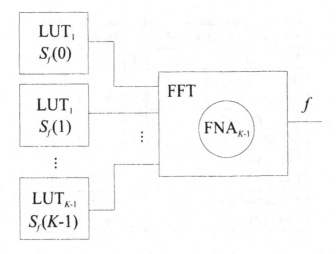

Fig. 4.57 Circuit realization of functions from FNAPDDs.

represented. Figure 4.56 shows this FANPDD with matrix-valued nodes represented by decision diagrams, which are reduced to a node with four outgoing edges for each submatrix. In this way, integer-valued FNAPDDs (ivFNAPDDs) are derived.

Complexities of representing f by using MTDD for $G_{24} = C_2 \times C_2 \times C_2 \times C_3$, and $G_{24} = C_4 \times C_6$ and FNAPDD is compared in Table 4.15.

Examples 4.14, 4.15, and 4.16 are all considering binary-valued functions where BDDs and MTDDs have a starting advantage of 2 constant nodes, since they are defined with respect to the trivial basis. In the general case, for arbitrary discrete functions the number of different Fourier coefficients is not greater than the number of different values a function can take. That means, FNADDs and FNAPDDs may have smaller number of constant nodes that MTDDs. That especially applies to the functions with so-called flat spectrum. FNAPDDs permits to represent a function through matrix-valued coefficients that can be recorded in a Look-up-table. It follows that FFT processors can be used to calculate the function values from spectral coefficients. The same approach can be used for circuit realizations of discrete functions. Figure 4.57 shows the basic principle of realization of functions through FNAPDDs.

4.13 ALGORITHM FOR CONSTRUCTION OF FNAPDD

In practical applications of FNAPDDs representations, we assume that a library of finite Abelian and non-Abelian groups and their representations is provided. The groups whose orders are compatible to the class of functions that can be represented are taken in this library. The orders of the selected groups and their representations should be factors in the orders of vectors representing the targeted class of functions. For example, the quaternion group Q_2 is convenient in representation of switching

Table 4.15 Complexities of representing f in Example 4.16 in terms of the number of levels, non-terminal nodes (ntn), constant nodes (cn), and sizes.

Number of	d	ntn	cn	s
$G_{24} = C_2 \times C_2 \times C_2 \times C_3$ MTDD	5	7	2	9
$G_{24} = C_4 \times C_6$ MTDD	3	5	2	7
Matrix-valued FNAPDD	2	1	3	4
Number-valued FNAPDD	4	8	8	16

functions, since has the order $8 = 2^3$ and the representations of order 2. Similar, the symmetric group of permutations S_3 is suitable for representation of three-valued and four-valued functions, since it is of order 6 with representations of order 2.

4.13.1 Algorithm for representation

For a given f, the FNAPDD representation is derived by using the following algorithm.

Algorithm 4.5 *(Determination of FNAPDD)*

1. *Given a function $f \in P(G)$. Determine a non-Abelian group G_n of order $g_n \leq g$ with representations of order r_w by using the Algorithm 6.*

2. *Transfer f into the corresponding matrix-valued function f_m by using the Algorithm A_2vm. f_m is represented by a vector $[\mathbf{F}_m]$ of order g_n whose elements are $(r_w \times r_w)$ matrices.*

3. *Represent f_m by the matrix-valued FNAPDD on G_n.*

4. *Represent the matrix-valued nodes in FNAPDD of f_m by DDs on groups of order r_w^2.*

5. *Do the possible optimizations of this representation.*

Algorithm 4.6 *(Generation of G_n)*

1. *Assume that f may be represented by a vector $\mathbf{G}$ of order g.*

2. *Read the orders of available groups g_i and their representations $\mathbf{R}_{w_i}$ in library of finite groups. Factorize g such that at least one factor of the form $r_{w_i}^2 g_i$ appears.*

3. *Choose the corresponding non-Abelian group of order g_i with representations of order r_{w_i} as a possible constituent subgroup G_i in G_n.*

4. *Repeat 2 if possible, otherwise choose the remaining constituent subgroups in G_n among the available Abelian groups. In that, do a reasonable compromise between the number of levels and number of outgoing edges of nodes associated to the chosen subgroups G_i in G_n.*

5. *Determine the structure of G_n as a group of order $g_n = \frac{g}{r_w}$ with representations of order $r_w = \prod_{i=1}^{k} r_{w_i}$, where k is determined in previous step and order the subgroups G_i chosen in 3 and 4 in the increasing order.*

6. *Determine the Fourier transform matrix $\mathbf{R}$ on G_n as the Kronecker product of the Fourier transform matrices on the constituent subgroups G_i.*

4.14 OPTIMIZATION OF FNAPDD

There are several possibilities for the optimization of FNAPDDs.

1. Choose the best suited group G_n depending on the possible factorizations of the order g of G on which f is initially given.

2. Do the optimization in generation of f_m from f by choosing between algorithms A_1 and A_2.

3. Do the optimization by choosing among DTs of various structures for the representation of the matrix-valued nodes in the matrix-valued FNAPDD. DDs on both Abelian or non-Abelian groups may be used.

4. Do the optimization in transferring the matrix-valued nodes into functions on the corresponding subgroups by choosing arbitrarily between the algorithms A_2mv by columns and by rows.

5. In representation of multi-output functions shared FNAPDDs may be used. These DDs are defined in the same way as Shared binary decision diagrams (SBDDs) and Shared multi-terminal binary decision diagrams (SMTBDDs) are defined [1].The optimization may be done as with SBDDs and SMTBDDs by ordering of the variables in generation the integer-valued counterpart $f(z)$ and by pairing of outputs.

REFERENCES

1. Babu, H.Md.H., Sasao, T., "A method to represent multiple-output switching functions by using binary decision diagrams", *SASIMI'96*, November 25, 1996.

2. Bernd, J., Gergov, J., Meinel, C., Slobodova, A., "Boolean manipulation with free BDDs, first experimental results", *Proc. European Design and Test Conf.*, February 28 - March 3, 1994, 200-207.

3. Bryant, R.E. "Graph-based algorithms for Boolean functions manipulation", *IEEE Trans. Comput.*, Vol. C-35, No. 8, 1986, 667-691.

4. Bryant, R.E., Chen, Y-A., "Verification of arithmetic functions with binary moment decision diagrams", May 31, 1994, CMU-CS-94-160.

5. Clarke, E.M., McMillan, K.L., Zhao, X., Fujita, M., "Spectral transforms for extremely large Boolean functions", in Kebschull, U., Schubert, E., Rosenstiel, W., Eds., *Proc. IFIP WG 10.5 Workshop on Applications of the Reed-Muller Expansion in Circuit Design*, 16-17.9.1993, Hamburg, Germany, 86-90.

6. Drechsler, R., Becker, B., OKFDDs-Algorithms, applications and extensions", in Sasao, T., Fujita, M., (Ed.), *Representations of Discrete Functions*, Kluwer Academic Publishers, Boston, 1996, 163-190.

7. Falkowski, B.J., Rahardja, S., "Complex spectral decision diagrams", *Proc. 26th Int. Symp. on Multiple-Valued Logic*, Santiago de Campostela, Spain, May 1996, 255-260.

8. Hengster, H., Drechsler, R., Eckrich, S., Pfeiffer, T., Becker, B., "AND/EXOR based synthesis of testable KFDD-circuits with small depth", *Proc. Asian Test Conf.*, 1996.

9. Karpovsky, M.G., "Fast Fourier transforms on finite non-Abelian groups", *IEEE Trans. Comput.*, Vol. C-26, No. 10, Oct. 1977, 1028-1030.

10. Karpovsky, M.G., Trachtenberg, E.A., "Some optimization problems for convolution systems over finite groups", *Inform. Control.*, Vol. 34, No. 3, 1977, 1-22.

11. Karpovsky, M.G., Trachtenberg, E.A., "Fourier transform over finite groups for error detection and error correction in computation channels", *Inform. Control*, Vol. 40, No. 3, 1979, 335-358.

12. Roziner, T.D., Karpovsky, M.G., Trachtenberg, L.A., "Fast Fourier transforms over finite groups by multiprocessor system", *IEEE Trans.*, Vol. ASSP-38, No. 2, 1990, 226-240.

13. Lai, Y.F., Pedram, M., Vrudhula, S.B.K., "EVBDD-based algorithms for integer linear programming, spectral transformation, and functional decomposition",

IEEE Trans. Computer-Aided Design of Integrated Circuits and Systems, Vol. 13, No. 8, 1994, 959-975.

14. Sasao, T., *Logic Design: Switching Circuit Theory*, Kindai Kaga-ku, Tokyo, Japan, 1995.

15. Sasao, T., *Switching Theory for Logic Synthesis*, Kluwer Academic Publishers, Boston, 1999.

16. Sasao, T., "Representations of logic functions by using EXOR operators", in Sasao, T., Fujita, M., (eds.), *Representations of Discrete Functions*, Kluwer Academic Publishers, Boston, 1996, 29-54.

17. Sasao, T., Fujita, M., *Representations of Discrete Functions*, Kluwer Academic Publishers, Boston, 1996.

18. Sasao, T., Butler, J.T., "A design method for look-up table type FPGA by pseudo-Kronecker expansions", *Proc. 24th Int. Symp. on Multiple-valued Logic*, Boston, May 25-27, 1994, 97-104.

19. Sasao, T., Butler, T.J., "A method to represent multiple-output functions by using multi-valued decision diagrams", *Proc. Int. Symp. on Multiple-Valued Logic*, May 29-31, 1996, Santiago de Campostela, Spain, 248-254.

20. Shannon, E.C., "The synthesis of two-terminal switching circuits", *Bell System Tech. J.*, Vol. 28, No. 1, 1949.

21. Stanković, R.S., "Some remarks about spectral transform interpretation of MTB-DDs and EVBDDs", *Proc. ASP-DAC'95*, August 29 - September 1, 1995, Makuhari Messe, Chiba, Japan, 385-390.

22. Stanković, R.S., "Fourier decision diagrams for optimization of decision diagrams representations of discrete functions", *Proc. Workshop on Post Binary-Ultra Large Scale Integration*, Santiago de Campostela, Spain, 1996, 8-12.

23. Stanković, R.S., "Simple theory of decision diagrams for representation of discrete functions", *Proc. Reed-Muller 2000*, Victoria, BC, Canada, May 20-21, 1999,

24. Stanković, R.S., Astola, J.T., "Design of decision diagrams with increased functionality of nodes through group theory", *IEICE Trans. Fundamentals*, Vol. E86-A, No. 3, 2003, 693-703.

25. Stanković, R.S., Astola, J.T., *Spectral Interpretation of Decision Diagrams*, Springer, New York, 2003.

26. Stanković, R.S., Astola, J.T., "Matrix-valued decision diagrams in representations of complex systems", *17th European Meeting on Cybernetics and Systems Research*, Vienna, Austria, April 13-16, 2004.

27. Stanković, R.S., Sasao, T., Moraga, C., "Spectral transform decision diagrams", in Sasao, T., Fujita, M., (eds.), *Representations of Discrete Functions*, Kluwer Academic Publishers, Boston, 1996, 55-92.

5

Functional Expressions on Quaternion Groups

In this chapter, we discuss arithmetic expressions on the quaternion groups [19] corresponding to the arithmetic expressions on C_2^n. The interest in arithmetic expressions on Q_2^r is motivated by the renewed recent interest in arithmetic expressions on C_2^n originating in their properties and applications

1. Parallelization of algorithms for large switching functions [9], [10];

2. Single expression for multi-output switching functions [14];

3. Some word-level decision diagrams are graphic representations of arithmetic expressions for discrete functions [17], [18], [20];

4. Different applications of arithmetic expressions such, as for example, checking of error probability in logic networks [9], [22].

We extended a method for derivation of polynomial expressions from Walsh (Fourier) expansions on C_2^n to non-Abelian groups by using as the example the group Q_2. For other values of n, we use the groups $Q_2^{n/3}$, $C_2 Q_2^{(n-1)/3}$, and $C_2^2 Q_2^{(n-2)/3}$. In this case, the arithmetic expressions are generated by the Kronecker product of the arithmetic matrices on C_2 and Q_2. We consider groups of orders 2^n, therefore, it is possible to derive arithmetic expressions on quaternion groups in terms of switching variables. Extension to groups of arbitrary orders is possible by allowing variables to take values in arbitrary finite sets. An example of such generalizations for groups Cp^n, for example $p = 3$, is given in [12].

5.1 FOURIER EXPRESSIONS ON FINITE DYADIC GROUPS

5.1.1 Finite dyadic groups

Group representations for C_2 are given by the basic Walsh matrix

$$\mathbf{W}(1) = \begin{bmatrix} 1 & 1 \\ 1 & -1 \end{bmatrix}.$$

The group representations of the finite dyadic group C_2^n are given by the Walsh matrix of order n defined as

$$\mathbf{W}(n) = \bigotimes_{i=1}^{n} \mathbf{W}(1),$$

where $\otimes$ denotes the Kronecker product.

We denote by $C(C_2^n)$ the space of functions $f : C_2^n \rightarrow C$, where C_2 is the cyclic group of order 2, and C is the complex-field. Assume that $f \in C(C_2^n)$ is given by the vector $\mathbf{F} = [f(0), \ldots, f(2^n - 1)]^T$.

If the Walsh (Fourier) spectrum for f is represented by a vector of spectral coefficients $\mathbf{S}_{w,f} = [S_{w,f}(0), \ldots, S_{w,f}(2^n - 1)]^T$, then

$$\mathbf{S}_{w,f} = 2^{-n}\mathbf{W}(n)\mathbf{F},$$

and

$$\mathbf{F} = \mathbf{W}(n)\mathbf{S}_{w,f},$$

since the Walsh matrix is a self-inverse matrix over C with the scaling factor 2^{-n}.

Example 5.1 *The Walsh spectrum of a three-variable switching function $f(x_1, x_2, x_3)$ given by the truth-vector $\mathbf{F} = [1, 0, 0, 0, 0, 1, 1, 1]^T$ is given by vector of coefficients $\mathbf{S}_{w,f} = [4, 0, 0, 0, -2, 2, 2, 2]^T$. Thus,*

$$f(x) = \frac{1}{8}(4w_0(x) - 2w_4(x) + 2w_5(x) + 2w_6(x) + 2w_7(x)), \tag{5.1}$$

where $w_i(x)$ represent the columns of the corresponding Walsh matrix.

5.2 FOURIER EXPRESSIONS ON Q_2

The Fourier transform on Q_2 is defined in terms of the group representations given by the columns of the matrix whose entries are determined from Table 2.9. From this table, the set of functions $R_w^{(i,j)}(x)$ used to define the Fourier transform on Q_2 is represented by the columns of the following matrix

$$
\mathbf{Q}_2 = \begin{bmatrix}
1 & 1 & 1 & 1 & 1 & 0 & 0 & 1 \\
1 & 1 & -1 & -1 & i & 0 & 0 & -i \\
1 & 1 & 1 & 1 & -1 & 0 & 0 & -1 \\
1 & 1 & -1 & -1 & -i & 0 & 0 & i \\
1 & -1 & 1 & -1 & 0 & -1 & 1 & 0 \\
1 & -1 & -1 & 1 & 0 & -i & -i & 0 \\
1 & -1 & 1 & -1 & 0 & 1 & -1 & 0 \\
1 & -1 & -1 & 1 & 0 & i & i & 0
\end{bmatrix}.
$$

Notice that in this matrix, the order of group representations is a bit different compared to Table 2.9, in particular, $\mathbf{R}_1$ and $\mathbf{R}_2$ are permuted.

The Fourier transform matrix is given by the matrix inverse to $\mathbf{Q}_2$. Thus,

$$
\mathbf{Q}_2^{-1} = \frac{1}{8} \begin{bmatrix}
1 & 1 & 1 & 1 & 1 & 1 & 1 & 1 \\
1 & 1 & 1 & 1 & -1 & -1 & -1 & -1 \\
1 & -1 & 1 & -1 & 1 & -1 & 1 & -1 \\
1 & -1 & 1 & -1 & -1 & 1 & -1 & 1 \\
2 & -2i & -2 & 2i & 0 & 0 & 0 & 0 \\
0 & 0 & 0 & 0 & -2 & 2i & 2 & -2i \\
0 & 0 & 0 & 0 & 2 & 2i & -2 & -2i \\
2 & 2i & -2 & -2i & 0 & 0 & 0 & 0
\end{bmatrix}.
$$

We denoted by $C(Q_2)$ the space of functions $f : Q_2 \to C$.
For a function $f \in C(Q_2)$ given by the vector

$$
\mathbf{F} = [f(0), f(1), f(2), f(3), f(4), f(5), f(6), f(7)]^T,
$$

the Fourier spectrum given in the matrix notation by the vector of spectral coefficients

$$
\mathbf{S}_{q,f} = [S_{q,f}(0), S_{q,f}(1), S_{q,f}(2), S_{q,f}(3), S_{q,f}(4), S_{q,f}(5), S_{q,f}(6), S_{q,f}(7)]^T,
$$

is determined as

$$
\mathbf{S}_{q,f} = \mathbf{Q}_2^{-1} \mathbf{F},
$$

and

$$
\mathbf{F} = \mathbf{Q}_2 \mathbf{S}_{q,f}.
$$

Example 5.2 *If f in Example 5.1 is considered as a function on Q_2, then the Fourier spectrum for f is given by $\mathbf{S}_{q,f} = \frac{1}{8}[4, -2, 0, 2, 2, 2 - 2, 2]^T$. Thus,*

$$
f(x) = \frac{1}{8}(4q_0(x) - 2q_1(x) + 2q_3(x) + 2q_4(x) + 2q_5(x) - 2q_6(x) + 2q_7(x)),
$$

where $q_i(x)$ are columns of the matrix $\mathbf{Q}_2$.

5.3 ARITHMETIC EXPRESSIONS

In the matrix notation, the arithmetic expressions for functions in $C(C_2^n)$ are defined as

$$f = \mathbf{X}_a(n)\mathbf{S}_{a,f}$$
$$= \left(\bigotimes_{i=1}^{n} [\ 1 \quad x_i\ ] \right) \mathbf{S}_{a,f}$$

with

$$\mathbf{S}_{a,f} = \mathbf{A}(n)\mathbf{F},$$

$\mathbf{F} = [f(0), \ldots, f(2^n - 1)]^T$, and

$$\mathbf{A}(n) = \bigotimes_{i=1}^{n} \mathbf{A}(1),$$

where $\mathbf{A}(1)$ is the basic arithmetic transform matrix given by

$$\mathbf{A}_f(1) = \begin{bmatrix} 1 & 0 \\ -1 & 1 \end{bmatrix}.$$

If the elements of $\mathbf{X}_a$ are interpreted as logic values 0 and 1, then this matrix is referred to as the Reed-Muller matrix or the conjunctive transform matrix [2],[3]. It defines a self-inverse transform in $GF_2(C_2^n)$ denoted as the Reed-Muller transform, or the conjunctive transform. In this context, the arithmetic transform in $C(C_2^n)$ is denoted as the inverse conjunctive transform [2]. For more details on arithmetic expressions, see for example, [2], [9], [10], [15], [22].

Example 5.3 *For $n = 3$, the arithmetic transform in $C(C_2^3)$ is defined by the matrix*

$$\mathbf{A}(3) = \begin{bmatrix}
1 & 0 & 0 & 0 & 0 & 0 & 0 & 0 \\
-1 & 1 & 0 & 0 & 0 & 0 & 0 & 0 \\
-1 & 0 & 1 & 0 & 0 & 0 & 0 & 0 \\
1 & -1 & -1 & 1 & 0 & 0 & 0 & 0 \\
-1 & 0 & 0 & 0 & 1 & 0 & 0 & 0 \\
1 & -1 & 0 & 0 & -1 & 1 & 0 & 0 \\
1 & -1 & 0 & 0 & -1 & 1 & 0 & 0 \\
1 & 0 & -1 & 0 & -1 & 0 & 1 & 0 \\
-1 & 1 & 1 & -1 & 1 & -1 & -1 & 1
\end{bmatrix}.$$

For f in Example 5.1, the arithmetic spectrum is given by

$$\mathbf{A}_f = [1, -1, -1, 1, -1, 2, 2, -2]^T.$$

Therefore,

$$f = 1 - x_3 - x_2 + x_2 x_3 - x_1 + 2x_1 x_3 + 2x_1 x_2 - 2x_1 x_2 x_3.$$

5.4 ARITHMETIC EXPRESSIONS FROM WALSH EXPANSIONS

Let $G = C_2^2$ and $P = C$. Each $x \in G$ can be expressed by $x = (x_1, x_2)$, $x_1, x_2 \in \{0, 1\}$. Fourier expansions for functions $f \in C(C_2^2)$ are defined in terms of Walsh functions given by the columns of Walsh matrix

$$\mathbf{W}(2) = \begin{bmatrix} 1 & 1 & 1 & 1 \\ 1 & -1 & 1 & -1 \\ 1 & 1 & -1 & -1 \\ 1 & -1 & -1 & 1 \end{bmatrix}. \tag{5.2}$$

The set of Walsh functions w_i, $i = 0, 1, 2, 3$ can be represented in terms of switching variables as

1. $w_0 = 1$,

2. $w_1 = 1 - 2x_2$,

3. $w_2 = 1 - 2x_1$,

4. $w_3 = (1 - 2x_1)(1 - 2x_2)$.

In symbolic notation, $\mathbf{W}(2)$ can be written as

$$\mathbf{X}(2) = \begin{bmatrix} w_0 & w_1 & w_2 & w_3 \end{bmatrix}.$$

From (5.2), the Walsh expression for $f \in C(C_2^2)$ is defined by

$$\begin{aligned} f &= \mathbf{X}(2)\mathbf{S}_{w,f} \\ &= \begin{bmatrix} 1 & 1 - 2x_2 & 1 - 2x_1 & (1 - 2x_1)(1 - 2x_2) \end{bmatrix} \mathbf{S}_f, \end{aligned}$$

where $\mathbf{S}_{w,f} = [S_{w,f}(0), S_{w,f}(1), S_{w,f}(2), S_{w,f}(3)]^T$ is the vector of Walsh spectral coefficients.

With this notation, the orthogonal Walsh series expression transfers into the Walsh polynomial expressions in terms of switching variables. It is assumed that switching variables are coded by $(0, 1)_{GF(2)} \rightarrow (0, 1)_Z$. Therefore,

$$\begin{aligned} f &= 1 \cdot S_{w,f}(0) + (1 - 2x_2)S_{w,f}(1) + (1 - 2x_1)S_{w,f}(2) \\ &\quad + (1 - 2x_1)(1 - 2x_2)S_{w,f}(3). \end{aligned} \tag{5.3}$$

In (5.3), if the multiplications are performed, then the polynomial expression for f is derived.

$$f = 1 \cdot z_0 - 2x_1 z_2 - 2x_2 z_1 + 4x_1 x_2 z_3, \tag{5.4}$$

where

$$\begin{aligned} z_0 &= S_{w,f}(0) + S_{w,f}(1) + S_{w,f}(2) + S_{w,f}(3), \\ z_1 &= S_{w,f}(1) + S_{w,f}(3), \\ z_2 &= S_{w,f}(2) + S_{w,f}(3), \\ z_3 &= S_{w,f}(3). \end{aligned}$$

Note that in this relation, indices of variables and coefficients are ordered in a way that corresponds to the Hadamard ordering of Walsh functions [8].

If Walsh (Fourier) coefficients are expressed in terms of the function values for f, these polynomial representations become the arithmetic expressions for f.

In this example,

$$S_{w,f}(0) = \frac{1}{4}(f(0) + f(1) + f(2) + f(3)),$$

$$S_{w,f}(1) = \frac{1}{4}(f(0) - f(1) + f(2) - f(3)),$$

$$S_{w,f}(2) = \frac{1}{4}(f(0) + f(1) - f(2) - f(3)),$$

$$S_{w,f}(3) = \frac{1}{4}(f(0) - f(1) - f(2) + f(3)).$$

Therefore,

$$z_0 = f(0),$$

$$z_1 = \frac{1}{2}(f(0) - f(1)),$$

$$z_2 = \frac{1}{2}(f(0) - f(2)),$$

$$z_3 = \frac{1}{4}(f(0) - f(1) - f(2) + f(3)).$$

After replacement of z_i in (5.4), we get the arithmetic expressions for $f \in C(C_2^n)$

$$f = 1 \cdot a_0 + x_1 a_2 + x_2 a_1 + x_1 x_2 a_3,$$

where

$$a_0 = f(0),$$
$$a_1 = f(0) - f(1),$$
$$a_2 = f(0) - f(2),$$
$$a_3 = f(0) - f(1) - f(2) + f(3).$$

If further

1. Binary values 0 and 1 for variables are considered as the logic values 0 and 1,

2. The addition and subtraction in C are replaced by the addition in $GF(2)$,

3. Values of coefficients are calculated modulo 2,

then, the arithmetic expressions become the Reed-Muller expressions for f.

In this example, the Reed-Muller expression is given by

$$f = 1 \cdot f_0 \oplus r_2 x_1 \oplus x_2 r_1 \oplus r_3 x_1 x_2,$$

where $r_i \in \{0, 1\}$, are

$$
\begin{aligned}
r_0 &= f(0), \\
r_1 &= f(0) \oplus f(1), \\
r_2 &= f(0) \oplus f(2), \\
r_3 &= f(0) \oplus f(1) \oplus f(2) \oplus f(3).
\end{aligned}
$$

Example 5.4 *For $n = 3$, the Walsh functions in the Hadamard ordering can be represented as*

$$
\begin{aligned}
w_0(x) &= 1, \\
w_1(x) &= 1 - 2x_3, \\
w_2(x) &= 1 - 2x_2, \\
w_3(x) &= (1 - 2x_2)(1 - x_3), \\
w_4(x) &= 1 - 2x_1, \\
w_5(x) &= (1 - 2x_1)(1 - 2x_3), \\
w_6(x) &= (1 - 2x_1)(1 - 2x_2), \\
w_7(x) &= (1 - 2x_1)(1 - 2x_2)(1 - 2x_3).
\end{aligned}
$$

If we replace these relations in (5.1), then

$$
\begin{aligned}
f &= \frac{1}{8}(4 - 2(1 - 2x_1) + 2(1 - 2x_1)(1 - 2x_3) \\
&\quad + 2(1 - 2x_1)(1 - 2x_2) + 2(1 - 2x_1)(1 - 2x_2)(1 - 2x_3)) \\
&= 1 - x_1 - x_2 - x_3 + 2x_1x_2 + 2x_1x_3 + x_2x_3 - 2x_1x_2x_3,
\end{aligned}
$$

which is the arithmetic expression for f.

If we reduce the coefficients modulo 2, and replace the addition and subtraction by EXOR, we get the Reed-Muller expression for f

$$
f = 1 \oplus x_1 \oplus x_2 \oplus x_3 \oplus x_2x_3.
$$

5.5 ARITHMETIC EXPRESSIONS ON Q_2

Let G be the Quaternion (non-Abelian) group Q_2. The columns of the matrix $\mathbf{Q}_2$ in terms of which the Fourier transform on Q_2 is defined can be represented by the following set of functions in terms of switching variables

1. $q_0 = 1$,

2. $q_1 = (1 - 2x_1)$,

3. $q_2 = (1 - 2x_3)$,

4. $q_3 = (1 - 2x_1)(1 - 2x_3)$,

5. $q_4 = \overline{x}_1(1 - 2x_2)(\overline{x}_3 + ix_3)$,

6. $q_5 = -x_1(1 - 2x_2)(\overline{x}_3 + ix_3)$,

7. $q_6 = x_1(1 - 2x_2)(\overline{x}_3 - ix_3)$,

8. $q_7 = \overline{x}_1(1 - 2x_2)(\overline{x}_3 - ix_3)$.

Each $f \in C(Q_2)$ can be represented as a linear combination of these functions. The coefficients in this expression are the Fourier coefficients on Q_2 over C.

Expressions,Arithmetic on non-Abelian groups

Example 5.5 *Consider a three-variable switching function f as a function on Q_2. If f is given by the truth-vector $\mathbf{F} = [1, 0, 0, 0, 0, 1, 1, 1]^T$, then the Fourier coefficients for f are given by the vector $\mathbf{S}_f = \frac{1}{2}[4, -2, 0, 2, 2, 2, -2, 2]^T$. Therefore,*

$$
\begin{aligned}
f \;=\; & \frac{1}{8}(4 - 2(1 - 2x_1) + 2(1 - 2x_1)(1 - 2x_2) \\
& +2\overline{x}_1(1 - 2x_2)(\overline{x}_3 + ix_3) - 2x_1(1 - 2x_2)(\overline{x}_3 + ix_3) \\
& -2x_1(1 - 2x_2)(\overline{x}_3 - ix_3) + 2x_1(1 - 2x_2)(\overline{x}_3 - ix_3)).
\end{aligned}
$$

Finally,

$$
f = \frac{1}{8}(4 - 4x_3 + 8x_1x_3 + 4\overline{x}_1\overline{x}_3 - 4x_1\overline{x}_3 - 8\overline{x}_1x_2\overline{x}_3 + 8x_1x_2\overline{x}_3). \qquad (5.5)
$$

In a general case,

$$
\begin{aligned}
f \;=\; & q_0 S_{q,f}(0) + q_1 S_{q,f}(1) + q_2 S_{q,f}(2) + q_3 S_{q,f}(3) \\
& +q_4 S_{q,f}(4) + q_5 S_{q,f}(5) + q_6 S_{q,f}(6) + q_7 S_{q,f}(7).
\end{aligned}
$$

From the matrix of the group representations on Q_2, the Fourier coefficients on Q_2 are given by

$$
\begin{aligned}
S_{q,f}(0) \;&=\; \frac{1}{8}(f(0) + f(1) + f(2) + f(3) + f(4) + f(5) + f(6) + f(7)) \\
S_{q,f}(1) \;&=\; \frac{1}{8}(f(0) + f(1) + f(2) + f(3) - f(4) - f(5) - f(6) - f(7)) \\
S_{q,f}(2) \;&=\; \frac{1}{8}(f(0) - f(1) + f(2) - f(3) + f(4) - f(5) + f(6) - f(7)) \\
S_{q,f}(3) \;&=\; \frac{1}{8}(f(0) - f(1) + f(2) - f(3) - f(4) + f(5) - f(6) + f(7)) \\
S_{q,f}(4) \;&=\; \frac{1}{8}(2f(0) - 2if(1) - 2f(2) + 2if(3)) \\
S_{q,f}(5) \;&=\; \frac{1}{8}(-2f(4) + 2if(5) + 2f(6) - 2if(7)) \\
S_{q,f}(6) \;&=\; \frac{1}{8}(2f(4) + 2if(5) - 2f(6) - 2if(7)) \\
S_{q,f}(7) \;&=\; \frac{1}{8}(2f(0) + 2if(1) - 2f(2) - 2if(3))
\end{aligned}
$$

From there,

$$
\begin{aligned}
f \;=\; & \frac{1}{2}(f(0) + f(2) + x_3(-f(0) + f(1) - f(2) + f(3)) \\
& + x_1(-f(0) - f(2) + f(4) + f(6)) \\
& + x_1 x_3(f(0) - f(1) + f(2) - f(3) - f(4) + f(5) - f(6) + f(7)) \\
& + \overline{x}_1 \overline{x}_3(1 - 2x_2)(f(0) - f(2)) \\
& \overline{x}_1 x_3(1 - 2x_2)(f(1) - f(3)) \\
& x_1 \overline{x}_3(1 - 2x_2)(f(4) - f(6)) \\
& x_1 x_3(1 - 2x_2)(f(5) - f(7)).
\end{aligned}
$$

In matrix notation, this expression can be considered as a series expansion in terms of the basic functions in $C(Q_2)$ represented by columns of the matrix

$$
\mathbf{X}_q =
\begin{bmatrix}
1 & 0 & 0 & 0 & 1 & 0 & 0 & 0 \\
1 & 1 & 0 & 0 & 0 & 1 & 0 & 0 \\
1 & 0 & 0 & 0 & -1 & 0 & 0 & 0 \\
1 & 1 & 0 & 0 & 0 & -1 & 0 & 0 \\
1 & 1 & 1 & 1 & 0 & 0 & 1 & 0 \\
1 & 0 & 1 & 0 & 0 & 0 & 0 & 1 \\
1 & 0 & 1 & 0 & 0 & 0 & -1 & 0 \\
1 & 1 & 1 & 1 & 0 & 0 & 0 & -1
\end{bmatrix}.
$$

If $f \in C(Q_2)$ is given by a vector $\mathbf{F} = [f(0), \ldots, f(7)]^T$, then the coefficients $\mathbf{Q}_f = [q_0, \ldots, q_7]^T$ in the expansion for f with respect to the basis given by $\mathbf{X}_q$ are determined as

$$
\mathbf{Q}_f = \mathbf{X}_q^{-1} \mathbf{F},
$$

where $\mathbf{X}_q^{-1}$ is the matrix inverse for $\mathbf{X}_q$ over C. Thus,

$$
\mathbf{X}_q^{-1} = \frac{1}{2}
\begin{bmatrix}
1 & 0 & 1 & 0 & 0 & 0 & 0 & 0 \\
-1 & 1 & -1 & 1 & 0 & 0 & 0 & 0 \\
-1 & 0 & -1 & 0 & 1 & 0 & 1 & 0 \\
1 & -1 & 1 & -1 & -1 & 1 & -1 & 1 \\
1 & 0 & -1 & 0 & 0 & 0 & 0 & 0 \\
0 & 1 & 0 & -1 & 0 & 0 & 0 & 0 \\
0 & 0 & 0 & 0 & 1 & 0 & -1 & 0 \\
0 & 0 & 0 & 0 & 0 & 1 & 0 & -1
\end{bmatrix}.
$$

This matrix can be considered as a transform matrix defined with respect to the basic functions represented by columns of the matrix $\mathbf{X}_q$. Thus, this matrix defines the inverse transform. We call this transform the Arithmetic-Haar transform, since the matrix $\mathbf{X}_q^{-1}$ exhibits a form similar to that of the Haar transform matrix [8].

Example 5.6 *For f in Example 5.1, the arithmetic-Haar coefficients are given by the vector* $\mathbf{S}_f = \frac{1}{2}[1, -1, 0, 2, 1, 0, -1, 0]^T$. *Therefore,*

$$f = \frac{1}{2}(1 - x_3 + 2x_1x_3 + \bar{x}_1\bar{x}_3(1 - 2x_2) - x_1\bar{x}_3(1 - 2x_2)).$$

Extensions of the arithmetic-Haar expressions from Q_2 to Q_2^n, $n = 3r$, can be done through the Kronecker product of the basic transform matrices on Q_2 in the same way as in the case of the arithmetic expressions and Walsh series expansions. Generalization to other values for n is easy, by using combinations of the basic arithmetic transform matrices. For example, if for a four variable switching function, the domain group is chosen as $G = C_2 \times Q_2$, then the arithmetic-Haar transform matrix is defined as the Kronecker product of $\mathbf{A}(1)$ and $\mathbf{X}_q$.

5.5.1 Arithmetic expressions and arithmetic-Haar expressions

In derivation of arithmetic-Haar expressions, we have exploited the correspondence between the order of C_2^n and $Q_2^{n/3}$. Due to that, we express the basic functions in arithmetic-Haar expressions in terms of switching variables, similar as that is done for the arithmetic expression on C_2^n. Therefore, it is natural to establish a relationship between the arithmetic expressions on C_2^n and the arithmetic-Haar expressions on $Q_2^{n/3}$.

The arithmetic-Haar expression for a given function f, converts into the arithmetic expression for f if we replace $\bar{x}_i \rightarrow (1 - x_i)$, and then recalculate the coefficients and assign them to the basic functions used in the arithmetic expressions.

Example 5.7 *For f in Example 5.6, the arithmetic-Haar expression is converted into the arithmetic expression as follows:*

$$
\begin{aligned}
f &= \frac{1}{2}(1 - x_3 + 2x_1x_3 + \bar{x}_1\bar{x}_3(1 - 2x_2) - x_1\bar{x}_3(1 - 2x_2)) \\
&= \frac{1}{2}(1 - x_3 + 2x_1x_3 + (1 - x_1)(1 - x_3)(1 - 2x_2) - x_1(1 - x_3)(1 - 2x_2)) \\
&= \frac{1}{2}(1 - x_3 + 2x_1x_3 + (1 - x_3)(1 - 2x_2) - 2x_1(1 - x_3)(1 - 2x_2)) \\
&= 1 - x_3 - x_2 + x_2x_3 - x_1 + 2x_1x_2 + 2x_1x_3 - 2x_1x_2x_3.
\end{aligned}
$$

5.5.2 Arithmetic-Haar expressions and Kronecker expressions

If we use a different representations for functions $q_0, \ldots, q_7$, we will derive different arithmetic expressions for functions in $C(Q_2)$. We use the above given representations, since they

  1. Resemble representations of Walsh functions in terms of switching variables,

  2. Provide the transform matrix similar to the Haar transform.

Arithmetic expressions in $C(C_2^3)$ are series expansions in terms of basic functions generated as

$$\begin{bmatrix} 1 & x_1 \end{bmatrix} \otimes \begin{bmatrix} 1 & x_2 \end{bmatrix} \otimes \begin{bmatrix} 1 & x_3 \end{bmatrix}.$$

If the values for variables are considered as the logic values, then the same basic functions are used in the definition of the Reed-Muller expressions. These basic functions can be generated by the recursive application of the positive Davio (pD) expansion defined as $f = 1 \cdot f_0 \oplus x_i(f_0 \oplus f_1)$, where $f_0 = f(x_i = 0)$, and $f_1 = f(x_i = 1)$.

Arithmetic-Haar expressions are series expansions in terms of the basic functions given by the symbolic matrix

$$\mathbf{X}_q = \begin{bmatrix} 1 & x_3 & x_1 & x_1 x_3 \\ \overline{x}_1 \overline{x}_3(1 - 2x_2) & \overline{x}_1 x_3(1 - 2x_2) & x_1 \overline{x}_3(1 - 2x_2) & x_1 x_3(1 - 2x_2) \end{bmatrix}.$$

These basic functions can be considered as a combination of two sets of basic functions in $C(C_2^3)$. First four basic functions are generated by the pD-expansion for variables x_1 and x_3. The other four basic functions are generated by the Shannon (S) expansion defined by $f = \overline{x}_i f_0 \oplus x_i f_1$ for x_1 and x_3, with multiplication by $(1 - 2x_2)$.

Thus, we generate the first four basic functions in the arithmetic-Haar expressions as

$$\begin{bmatrix} 1 & x_1 \end{bmatrix} \otimes \begin{bmatrix} 1 & x_3 \end{bmatrix} = \begin{bmatrix} 1 & x_3 & x_1 & x_1 x_3 \end{bmatrix}.$$

For other four basic functions we first use the S-expansion and get

$$\begin{bmatrix} \overline{x}_1 & x_1 \end{bmatrix} \otimes \begin{bmatrix} \overline{x}_3 & x_3 \end{bmatrix} = \begin{bmatrix} \overline{x}_1 \overline{x}_3 & \overline{x}_1 x_3 & x_1 \overline{x}_3 & x_1 x_3 \end{bmatrix}.$$

Then, we multiply the generated terms by $(1 - 2x_2)$.

For functions in $GF_2(C_2^n)$, expressions generated by combination of pD-expansions and S-expansions are denoted as Kronecker expressions [13]. In this setting, the arithmetic-Haar expressions are modified integer counterparts of the Kronecker expressions for switching functions.

5.6 DIFFERENT POLARITY POLYNOMIAL EXPRESSIONS

Switching variables take values in C_2. Negative literals of switching variables are defined as $\overline{x} = x \oplus 1$. The fixed-polarity Reed-Muller expressions are defined by freely choosing between x and $\overline{x}$ for each variable in f [14]. It is assumed that a particular variable appears as the positive or the negative literal, but not both in the same expressions for f. For a given f, fixed-polarity Reed-Muller expressions differ in the number of non-zero coefficients.

The polarity of variables chosen in an expression is conveniently expressed by the polarity vector $H = (h_1, \ldots, h_n)$, $h_i \in \{0, 1\}$. It is assumed that if $h_i = 1$, then the

i-th variable is represented by the negated literal $\overline{x_i}$. Thus, $h_i = 0$, shows that the i-th variable appears as x_i. The polarity where $h_i = 0$ for each $i = 1, \ldots, n$ is denoted as the zero-polarity. For a given class of expressions and a given f, the polarity for which the expression has the minimal number of non-zero coefficients is denoted as the optimal polarity.

In Fourier expansions for functions in $C(Q_2)$, the basic functions can be expressed in terms of switching variables. Thanks to that, the same method can be used to define fixed-polarity Fourier expansions $C(Q_2)$, and the arithmetic-Haar expression for functions in $C(C_2^n)$ and $C(Q_2^{n/3})$, $C_2 Q_2^{(n-1)/3}$, and $C_2^2 Q_2^{(n-2)/3}$.

5.6.1   Fixed-polarity Fourier expansions in $C(Q_2)$

Fixed-polarity Fourier expansions on Q_2 will be introduced through the example for $H = (101)$.

Example 5.8 *The use of negated literals for variables x_1 and x_3 transfers the basic functions for the Fourier transform on Q_2 into*

1. $q_0^{(101)} = 1$,

2. $q_1^{(101)} = 1 - 2\overline{x}_1$,

3. $q_2^{(101)} = 1 - 2\overline{x}_3$,

4. $q_3^{(101)} = (1 - 2\overline{x}_1)(1 - 2\overline{x}_3)$,

5. $q_4^{(101)} = x_1(1 - 2x_2)(x_3 + i\overline{x}_3)$,

6. $q_5^{(101)} = -\overline{x}_1(1 - 2x_2)(x_3 + i\overline{x}_3)$,

7. $q_6^{(101)} = \overline{x}_1(1 - 2x_2)(x_3 - i\overline{x}_3)$,

8. $q_7^{(101)} = x_1(1 - 2x_2)(x_3 - i\overline{x}_3)$.

Thus, the matrix of basic functions $\mathbf{Q}_2$ *for the Fourier transform in* $C(Q_2)$ *for the polarity* $H = (101)$ *is given by*

$$
\mathbf{Q}_8^{(101)} =
\begin{bmatrix}
1 & -1 & -1 & 1 & 0 & i & -i & 0 \\
1 & -1 & 1 & -1 & 0 & 1 & 1 & 0 \\
1 & -1 & -1 & 1 & 0 & -i & i & 0 \\
1 & -1 & 1 & -1 & 0 & -1 & -1 & 0 \\
1 & 1 & -1 & -1 & i & 0 & 0 & -i \\
1 & 1 & 1 & 1 & 1 & 0 & 0 & 1 \\
1 & 1 & -1 & -1 & -i & 0 & 0 & i \\
1 & 1 & 1 & 1 & -1 & 0 & 0 & -1
\end{bmatrix}.
$$

It follows that the columns of $\mathbf{Q}_2^{(101)}$ are given by $\mathbf{q}_0$, $-\mathbf{q}_1$, $-\mathbf{q}_2$, $\mathbf{q}_3$, $i\mathbf{q}_6$, $i\mathbf{q}_7$, $-i\mathbf{q}_4$, $i\mathbf{q}_5$. This permutation and change of columns in $\mathbf{Q}_2^{(101)}$ requires a permutation and change of rows in $\mathbf{Q}_2^{-1}$. We first transform rows in $\mathbf{Q}_2^{-1}$ in the same way as we transform the columns in $\mathbf{Q}_2$ to get $\mathbf{Q}_2^{(101)}$. In this way we get

$$
(\mathbf{Q}_8^{-1})' =
\begin{bmatrix}
1 & 1 & 1 & 1 & 1 & 1 & 1 & 1 \\
-1 & -1 & -1 & -1 & 1 & 1 & 1 & 1 \\
-1 & 1 & -1 & 1 & -1 & 1 & -1 & 1 \\
1 & -1 & 1 & -1 & -1 & 1 & -1 & 1 \\
0 & 0 & 0 & 0 & 2i & -2 & -2i & 2 \\
2i & -2 & -2i & 2 & 0 & 0 & 0 & 0 \\
-2i & -2 & 2i & 2 & 0 & 0 & 0 & 0 \\
0 & 0 & 0 & 0 & -2i & -2 & 2i & 2
\end{bmatrix}.
$$

Then, we permute the columns of $(\mathbf{Q}_2^{-1})'$ as is determined by the rule $x_1 x_2 x_3 \oplus \overline{h}_1 \overline{h}_2 \overline{h}_3$. Therefore, the columns of $(\mathbf{Q}_2^{-1})'$ are ordered as q_2', q_3', q_0', q_1', q_6', q_7', q_4', q_5'. In this way, we get the matrix $(\mathbf{Q}_2^{(101)})^{-1}$ inverse to $\mathbf{Q}_2^{(101)}$. It is given by

$$
(\mathbf{Q}_2^{(101)})^{-1} = \frac{1}{8}
\begin{bmatrix}
1 & 1 & 1 & 1 & 1 & 1 & 1 & 1 \\
-1 & -1 & -1 & -1 & 1 & 1 & 1 & 1 \\
-1 & 1 & -1 & 1 & -1 & 1 & -1 & 1 \\
1 & -1 & 1 & -1 & -1 & 1 & -1 & 1 \\
0 & 0 & 0 & 0 & -2i & 2 & 2i & -2 \\
-2i & 2 & 2i & -2 & 0 & 0 & 0 & 0 \\
2i & 2 & -2i & -2 & 0 & 0 & 0 & 0 \\
0 & 0 & 0 & 0 & 2i & 2 & -2i & -2
\end{bmatrix}.
$$

For f in Example 5.1, the Fourier spectrum for the polarity $H = (101)$ is given by

$$
\mathbf{S}_{q,f}^{(101)} = \frac{1}{8}[4, 2, 0, 2, 2i, -2i, 2i, -2i]^T.
$$

Therefore,

$$
\begin{aligned}
f = \ & \frac{1}{8}(4 + 2(1 - 2\overline{x}_1) + 2(1 - 2\overline{x}_1)(1 - 2\overline{x}_3) + 2ix_1(1 - 2x_2)(x_3 + i\overline{x}_3) \\
& - 2i\overline{x}_1(1 - 2x_2)(x_3 + i\overline{x}_3) + 2i\overline{x}_1(1 - 2x_2)(x_3 - i\overline{x}_3) \\
& - 2ix_1(1 - 2x_2)(x_3 - i\overline{x}_3)).
\end{aligned}
$$

5.6.2 Fixed-polarity arithmetic-Haar expressions

The matrix of basic arithmetic-Haar functions does not have complex values, and therefore, the transformation of this matrix into the matrix for the given polarity H is simpler than in the case of the Fourier transform.

We will introduce the fixed-polarity arithmetic-Haar expressions in $C(Q_2)$ with the example for $H = (101)$.

Example 5.9 *For $H = (101)$, the basic function for the arithmetic-Haar expressions in $C(Q_2)$ are given by columns of the matrix*

$$\mathbf{X}_q^{(101)} = \begin{bmatrix} 1 & 0 & 1 & 0 & 0 & 0 & 0 & 1 \\ 1 & 1 & 1 & 1 & 0 & 0 & 1 & 0 \\ 1 & 1 & 1 & 1 & 0 & 0 & 0 & -1 \\ 1 & 0 & 1 & 0 & 0 & 0 & -1 & 0 \\ 1 & 1 & 0 & 0 & 0 & 1 & 0 & 0 \\ 1 & 0 & 0 & 0 & 1 & 0 & 0 & 0 \\ 1 & 1 & 0 & 0 & 0 & -1 & 0 & 0 \\ 1 & 0 & 0 & 0 & -1 & 0 & 0 & 0 \end{bmatrix}.$$

This matrix is obtained by the permutation of rows in $\mathbf{X}_q$ by the rule $x_1 x_2 x_3 \oplus h_1 h_2 h_3$. Therefore, the rows with indices 0,1,2,3,4,5,6,7 are permuted as 5,4,7,6,1,0,3,2. This permutation requires the same permutation of columns in $\mathbf{X}_q^{-1}$. Therefore,

$$(\mathbf{X}_q^{(101)})^{-1} = \frac{1}{2} \begin{bmatrix} 0 & 0 & 0 & 0 & 0 & 1 & 0 & 1 \\ 0 & 0 & 0 & 0 & 1 & -1 & 1 & -1 \\ 0 & 1 & 0 & 1 & 0 & -1 & 0 & -1 \\ 1 & -1 & 1 & -1 & -1 & 1 & -1 & 1 \\ 0 & 0 & 0 & 0 & 0 & 1 & 0 & -1 \\ 0 & 0 & 0 & 0 & 1 & 0 & -1 & 0 \\ 0 & 1 & 0 & -1 & 0 & 0 & 0 & 0 \\ 1 & 0 & -1 & 0 & 0 & 0 & 0 & 0 \end{bmatrix}.$$

For f in Example 5.1, the arithmetic-Haar expression for the polarity $H = (101)$ is given by

$$f = \frac{1}{2}\left(2 - \overline{x}_3 - 2\overline{x}_1 + 2\overline{x}_1\overline{x}_3 - x_1\overline{x}_3(1 - 2x_2) + \overline{x}_1\overline{x}_3(1 - 2x_2)\right).$$

This expression requires six coefficients instead of five coefficients in the zero-polarity arithmetic-Haar expression for f.

Table 5.1 shows different spectra for f in Example 5.1, for the zero-polarity and the polarity $H = (101)$. Table 5.2 shows the corresponding expressions for f.

Table 5.3 compares the number of non-zero coefficients in some $mcnc$ benchmark functions used in logic design for the Walsh (W), the arithmetic (A), the Fourier on Q_2 (F), and the arithmetic-Haar (AH) transform. For multiple-output functions, each output is considered as a separate switching function, and in this table, fun-i denotes the i-th output of fun. This table shows that for each transform we can find a function where this transform provides a spectrum with fewer number of non-zero coefficients. In the most cases, the arithmetic and the Walsh transform are the most efficient, since are applied to binary-valued functions. However, even the complex-valued Fourier transform although applied to binary-valued functions can produce simpler spectrum than these transforms. For example, for sao2-i, $i = 1, 2, 3, 4$, the Fourier transform is more convenient than both the arithmetic and the Walsh

Table 5.1 Different spectra for f.

Expression	Spectrum
RM	$[1,1,1,1,1,0,0,0,]^T$
$RM^{(101)}$	$[1,1,0,1,1,0,0,1]^T$
A	$[1,-1,-1,1,-1,2,2,-2]^T$
$A^{(101)}$	$[1,-1,0,1,-1,2,0,-2]^T$
AH	$\frac{1}{2}[1,-1,0,2,1,0,-1,0]^T$
$AH^{(101)}$	$\frac{1}{2}[2,-1,-2,2,0,-1,0,1]^T$
W	$\frac{1}{8}[4,0,0,0,-2,2,2,2]^T$
$W^{(101)}$	$\frac{1}{8}[4,0,0,0,2,2,-2,2]^T$
F	$\frac{1}{8}[4,-2,0,2,2,2,-2,2]^T$
$F^{(101)}$	$\frac{1}{8}[4,2,0,2,2i,-2i,2i,-2i]^T$

Table 5.2 Different expressions for f.

RM	$f = 1 \oplus x_3 \oplus x_2 \oplus x_2 x_3 \oplus x_1$
$RM^{(101)}$	$f = 1 \oplus \overline{x}_3 \oplus x_2 \overline{x}_3 \oplus \overline{x}_1 \oplus \overline{x}_1 x_2 \overline{x}_3$
A	$f = 1 - x_3 - x_2 + x_2 x_3 - x_1 + 2x_1 x_2 + 2x_1 x_3 - 2x_1 x_2 x_3$
$A^{(101)}$	$f = 1 - \overline{x}_3 + x_2 \overline{x}_3 - \overline{x}_1 + 2\overline{x}_1 \overline{x}_3 - 2\overline{x}_1 x_2 \overline{x}_3$
AH	$f = \frac{1}{2}(1 - x_3 + 2x_1 x_3 + \overline{x}_1 \overline{x}_3(1 - 2x_2) - x_1 \overline{x}_3(1 - 2x_2))$
$AH^{(101)}$	$f = \frac{1}{2}(2 - \overline{x}_3 - 2\overline{x}_1 + 2\overline{x}_1 \overline{x}_3 - x_1 \overline{x}_3(1 - 2x_2) + \overline{x}_1 \overline{x}_3(1 - 2x_2))$
W	$f = \frac{1}{8}(4 - 2w_0(x) + 2w_5(x) + 2w_6(x) + 2w_7(x)), \ x = (x_1 x_2 x_3)$
$W^{(101)}$	$f = \frac{1}{8}(4 + 2(1 - 2\overline{x}_1) + 2(1 - 2\overline{x}_1)(1 - 2\overline{x}_3)$
	$-2(1 - 2\overline{x}_1)(1 - 2x_2)(1 - 2\overline{x}_3)$
F	$f = \frac{1}{8}(4 - 2(1 - 2x_1) + 2(1 - 2x_1)(1 - 2x_2)$
	$+2\overline{x}_1(1 - 2x_2)(\overline{x}_3 + ix_3)$
	$+2x_1(1 - 2x_2)(\overline{x}_3 + ix_3) - 2x_1(1 - 2x_2)(\overline{x}_3 - ix_3)$
	$+2\overline{x}_1(1 - 2x_2)(\overline{x}_3 - ix_3))$
$F^{(101)}$	$f = \frac{1}{8}(4 + 2(1 - 2\overline{x}_1) + 2(1 - 2\overline{x}_1)(1 - 2\overline{x}_3)$
	$+2ix_1(1 - 2x_2)(x_3 + i\overline{x}_3) - 2i\overline{x}_1(1 - 2x_2)(x_3 + i\overline{x}_3)$
	$+2i\overline{x}_1(1 - 2x_2)(x_3 - i\overline{x}_3) - 2ix_1(1 - 2x_2)(x_3 - i\overline{x}_3))$

transform. However, in this case, the arithmetic-Haar transform further reduces the number of non-zero coefficients. Conversely, for a given function, we should try different transforms to determine a spectrum with minimum number of non-zero coefficients. For example, for rd84-2 and rd84-3, the Fourier transform produces spectra with fewer non-zero coefficients than the arithmetic and the Walsh transform, respectively. For rd84-3, the arithmetic-Haar spectrum is comparable to the most efficient transform. For 9sym, the arithmetic-Haar transform requires fewer non-zero coefficients than the arithmetic transform, and the Walsh transform is the most efficient. For 5xp1-2, the arithmetic-Haar transform is more efficient than the Walsh transform, and the arithmetic transform is the most efficient. However, for 5xp1-3, the arithmetic-Haar transform is the most efficient. For all the outputs of sao2, the Fourier transform requires fewer products than either Walsh or the arithmetic transform, and the arithmetic-Haar transform is the most efficient.

We consider this analysis as a justification for introduction and use of different transforms, supported by a requirement that these transforms may share some useful properties of existing transforms. In that respect, the arithmetic-Haar transform is derived from the Fourier transform on Q_8^r in a uniform way as the arithmetic transform is derived from the Walsh transform on C_2^n. It eliminates the complex values in the Fourier transform and the corresponding transform matrix expresses a structure similar to the structure of the Haar matrix.

Example 5.10 *Assume the 16bits binary representation for integers x and y. Then, the function $y = 2^{16} sin(i/2^{16})$ can be considered as 16-variable 16-output switching function f. If we perform the addition of outputs multiplied with weighting coefficients 2^i, $i = 0, \ldots, 15$, we determine an integer-valued equivalent function f_Z for f. Table 5.4 shows the number of non-zero coefficients for y in the Fourier spectrum and the arithmetic-Haar spectrum on $C_2 Q_2^5$, and the Walsh spectrum, the arithmetic spectrum and the Haar spectrum on C_2^{16}. It is interesting to note that the complex-valued Fourier spectrum requires the fewer number of non-zero coefficients. Among them 5248 are real, the 5120 are the imaginary, and 7776 are the complex-valued coefficients.*

5.7   CALCULATION OF THE ARITHMETIC-HAAR COEFFICIENTS

5.7.1   FFT-like algorithm

The arithmetic-Haar transform matrix $\mathbf{X}_q$ has a form similar to that of the Haar transform matrix. Thanks to that property, it is possible to derive a FFT-like algorithm for calculation of the arithmetic-Haar coefficients. This algorithm is based upon the

Table 5.3 Number of non-zero coefficients.

f	F	AH	W	A
5xp1-1	96	26	80	18
5xp1-2	122	50	128	33
5xp1-3	124	34	65	41
5xp1-4	70	56	33	27
5xp1-5	84	36	17	17
5xp1-6	64	44	9	9
5xp1-7	56	32	5	5
5xp1-8	28	20	2	3
5xp1-9	24	8	2	2
5xp1-10	64	10	128	7
9sym	464	300	256	465
rd84-1	220	189	256	191
rd84-2	160	52	2	255
rd84-3	48	4	256	1
rd84-4	237	166	256	163
sao2-1	202	200	1024	380
sao2-2	275	269	640	768
sao2-3	348	297	856	578
sao2-4	322	216	1024	936

following factorization of $\mathbf{X}_q^{-1}$

$$
\mathbf{X}_q^{-1} =
\begin{bmatrix}
1 & 0 & 0 & 0 & 0 & 0 & 0 & 0 \\
0 & 1 & 0 & 0 & 0 & 0 & 0 & 0 \\
1 & 0 & 0 & 0 & 1 & 0 & 0 & 0 \\
0 & 1 & 0 & 0 & 0 & 1 & 0 & 0 \\
0 & 0 & 1 & 0 & 0 & 0 & 0 & 0 \\
0 & 0 & 0 & 1 & 0 & 0 & 0 & 0 \\
0 & 0 & 0 & 0 & 0 & 0 & 1 & 0 \\
0 & 0 & 0 & 0 & 0 & 0 & 0 & 1
\end{bmatrix}
\begin{bmatrix}
1 & 0 & 0 & 0 & 0 & 0 & 0 & 0 \\
-1 & 1 & 0 & 0 & 0 & 0 & 0 & 0 \\
0 & 0 & 1 & 0 & 0 & 0 & 0 & 0 \\
0 & 0 & 0 & 1 & 0 & 0 & 0 & 0 \\
0 & 0 & 0 & 0 & 1 & 0 & 0 & 0 \\
0 & 0 & 0 & 0 & -1 & 1 & 0 & 0 \\
0 & 0 & 0 & 0 & 0 & 0 & 1 & 0 \\
0 & 0 & 0 & 0 & 0 & 0 & 0 & 1
\end{bmatrix}
$$
$$
\begin{bmatrix}
1 & 0 & 1 & 0 & 0 & 0 & 0 & 0 \\
0 & 1 & 0 & 1 & 0 & 0 & 0 & 0 \\
1 & 0 & -1 & 0 & 0 & 0 & 0 & 0 \\
0 & 1 & 0 & -1 & 0 & 0 & 0 & 0 \\
0 & 0 & 0 & 0 & 1 & 0 & 1 & 0 \\
0 & 0 & 0 & 0 & 0 & 1 & 0 & 1 \\
0 & 0 & 0 & 0 & 1 & 0 & -1 & 0 \\
0 & 0 & 0 & 0 & 0 & 1 & 0 & -1
\end{bmatrix}.
$$

Figure 5.1 compares the flow-graphs of the fast algorithms for calculation of the Walsh, the arithmetic, the Fourier, and the arithmetic-Haar coefficients for functions in $f \in C(C_2^3$, and $f \in C(Q_2)$.

5.7.2   Calculation of arithmetic-Haar coefficients through decision diagrams

Decision diagrams (DDs) are an efficient data structure for compact representation of discrete functions [1], [4], [14]. Among many other applications, DDs are efficiently used to calculate different spectral transforms. From spectral interpretation of DDs [20], it was easy to show that in calculation of spectral transforms through DDs we actually perform basic operations in FFT-like algorithms. However, in DDs, calculation of the spectrum for f is not based on the vector of function values, but is performed over the DD for f. That permits to use the particular properties of f, which provides the computation efficiency and possibility to process large functions [5], [14], [21].

Calculation of the coefficients in arithmetic expressions of switching functions through DDs was considered in several papers, see for example [6], [7], [11].

In this section, we present a method to calculate arithmetic-Haar coefficients through decision diagrams (DDs) [1], [4]. The method is derived as a suitable modification of the corresponding method for calculation of the Haar spectrum [16].

We assume that a given $f \in C(Q_2^r)$ is alternatively considered as a function in $C(C_2^n)$, $n = 3r$, and represented by a Binary DD [1], [4], or a Multi-terminal binary DD (MTBDD) [5], depending on the function values taken in $GF(2)$ or C, respectively.

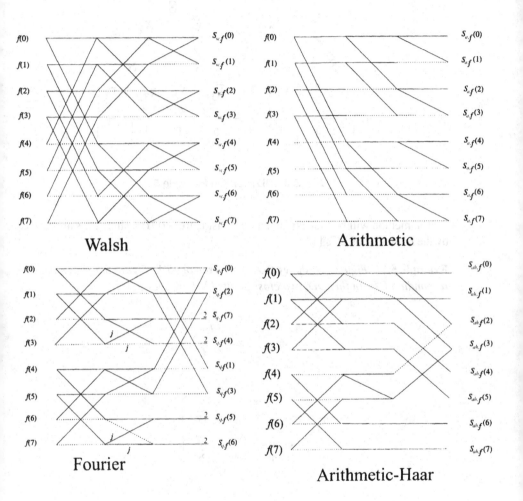

Fig. 5.1 FFT-like algorithms for $n = 3$.

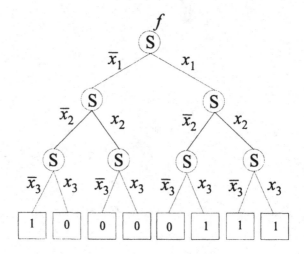

Fig. 5.2   BDT for f in Example 5.1.

The method will be first explained on decision trees (DTs), since DDs are derived by the reduction of DTs [14].

Example 5.11 *Figure 5.2 shows BDT f in Example 5.1. Figure 5.3 shows the corresponding DD. In this DD, two cross points [20] are shown.*

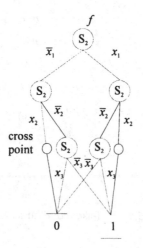

Fig. 5.3   BDD for F in Example 5.1.

We will present the method in a general form for functions in $C(Q_2^n)$, but we illustrate the method by the example of functions $f \in C(Q_2)$.

We assign to each node two fields, where we write the result of calculation through DDs. Calculations consist of processing the nodes in the DT for f. In the case of DDs, cross points [20] should be also processed.

In a DD, the leftmost node at a level is the node which can be reached by the path labeled by a product of variables consisting of negated literals.

In a DT, values of constant nodes are function values for f. We assume that at the level $(n-1)$ in the DT for f, the paths $\overline{x}_{n-1}x_n$ and $x_{n-1}\overline{x}_n$ are permuted. Thus, in the vector $\mathbf{F}$ representing f, the pairs of adjacent elements, starting from $f(1)$ are permuted.

For $n = 3$, f is given by $\mathbf{F} = [f(0), f(1), f(2), f(3), f(4), f(5), f(6), f(7)]^T$. After the mentioned reordering, the values of constant nodes in the DT for f are ordered as $\mathbf{F}_q = [f(0), f(2), f(1), f(3), f(4), f(6), f(5), f(7)]^T$.

The nodes at the level x_n are processed by using the rules described by rows of the matrix $\mathbf{W}(1)$. The values of the left and the right fields are determined by using the first row and the second row of $\mathbf{W}(1)$, respectively. The values of both fields for the nodes at all other levels are determined by using the rule described by the matrix $[-1, 1]$. The input data for calculation of values of left and right fields are determined as follows.

At the i-th level in the DT, if the node is pointed by the edge labeled with the negated literal $\overline{x}_i$, the input data for the left and right fields are reached by the subpaths $\overline{x}_i\overline{x}_{i+1}$ and $\overline{x}_i x_{i+1}$, respectively. If the node is pointed by the edge labeled with the positive literal, then the input data for the left and right fields are reached by the subpaths $\overline{x}_i\overline{x}_{i+1}$ and $x_i\overline{x}_{i+1}$, respectively.

The coefficient $S_{a,f}(0)$ is shown in the left field assigned to the leftmost node at the level x_n. The values of left fields assigned to the other nodes at this level are used in further calculations. The right fields of nodes at the level x_n show values of arithmetic-Haar coefficients $S_{a,f}(2^{n-1})$ to $S_{a,f}(2^n - 1)$. *For $n = 3$, these are coefficients $S_{a,f}(4)$, $S_{a,f}(5)$, $S_{a,f}(6)$, and $S_{a,f}(7)$.*

The other coefficients are shown in the left and right fields of the leftmost nodes from the level $n - 1$ to the root node. These coefficients are shown in the increasing order. Thus, the leftmost node at the level x_{n-1} shows $S_{a,f}(1)$, while the right field of this node shows $S_{a,f}(2)$. The root node shows the coefficient $S_{a,f}(2^{n-1} - 1)$. *For $n = 3$, the leftmost nodes at level x_2 shows the values of $S_{a,f}(1)$ and $S_{a,f}(2)$. The root node shows $S_{a,f}(3)$.*

Example 5.12 *Figure 5.4 shows calculation of arithmetic-Haar coefficients for $f \in C(Q_2)$ though DT. Figure 5.5 shows calculation of arithmetic-Haar coefficients for f in Example 5.2 through BDT. Figure 5.6 shows the same calculation through BDD. Note that in this example, permutation of paths $\overline{x}_2 x_3$ and $x_2\overline{x}_3$ does not permute values of constant nodes, since $f(1) = f(2) = 0$, $f(3) = f(4) = 0$, and $f(5) = f(6) = 1$.*

This algorithm can be used to calculate arithmetic-Haar coefficients of complex-valued functions if instead of BDDs, we use MTBDDs.

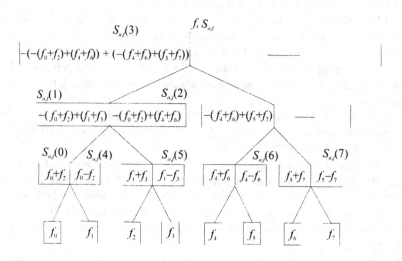

Fig. 5.4 Calculation of arithmetic-Haar coefficients through BDT.

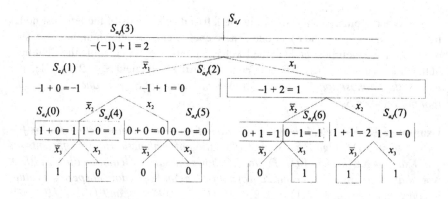

Fig. 5.5 Calculation of arithmetic-Haar coefficients for f in Example 5.2 through BDT.

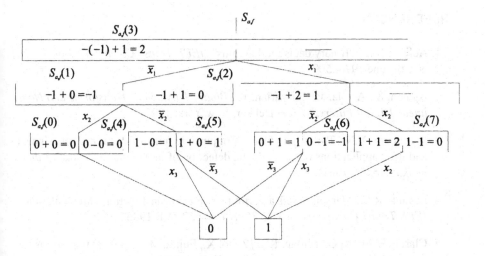

Fig. 5.6 Calculation of arithmetic-Haar coefficients for f in Example 5.2 through BDD.

REFERENCES

1. Ackers, S.B., "Binary decision diagrams", *IEEE Trans. Computers*, Vol. C-27, No. 6, June 1978, 509-516.

2. Agaian, S., Astola, J., Egiazarian, K., *Binary Polynomial Transforms and Nonlinear Digital Filters*, Marcel Dekker, New York, 1995.

3. Aizenberg, N.N., Trofimljuk, O.T., "Conjunctive transforms for discrete signals and their applications of tests and the detection of monotone functions", *Kibernetika*, No. 5, (1981).

4. Bryant, R.E., "Graph-based algorithms for Boolean functions manipulation", *IEEE Trans. Computers*, Vol. C-35, No. 8, 667-691, 1986.

5. Clarke, E, M., M.C., Millan, K.L., Zhao, X., Fujita, M., "Spectral transforms for extremely large Boolean functions", in Kebschull, U., Schubert, E., Rosenstiel, W., Eds., *Proc. IFIP WG 10.5 Workshop on Applications of the Reed-Muller Expression in Circuit Design*, Hamburg, Germany, September 16-17, 1993, 86-90.

6. Falkowski, B.J., Chang, C.H., "Efficient algorithm for the calculation of arithmetic spectrum from OBDD and synthesis of OBDD from arithmetic spectrum for incompletely specified Boolean functions", *Proc. 27th IEEE Int. Symp. on Circuits and Systems ISCAS94*, London, Vol. 1, May 1994, 197-200.

7. Falkowski, B.J., Chang, C.H., "Calculation of arithmetic spectra from free binary decision diagrams", *Proc. IEEE Int. Symp. on Circuits and Systems* (30th ISCAS), Hong Kong, June 1997, 1764-1767.

8. Karpovsky, M.G., *Finite Orthogonal Series in the Design of Digital Devices*, Wiley and JUP, New York and Jerusalem, 1976.

9. Kukharev, G.A., Shmerko, V.P., Yanushkievich, S.N., *Technique of Binary Data Parallel Processing for VLSI*, Vyshaja shcola, Minsk, Belarus, 1991.

10. Malyugin, V.D., *Paralleled Calculation by Means of Arithmetic Polynomials*, Physical and Mathematical Publishing Company, Russian Academy of Science, Moscow, 1997.

11. Malyugin, V.D., Stanković, R.S., Stanković, M., "Calculations of the coefficients of polynomial representations of switching functions through binary decision diagrams", *Proc. Preventive Engineering and Information Technologies*, Niš, Yugoslavia, December 8-12, 1994, 10-1-10-4.

12. Moraga, C., "On some applications of the Chrestenson functions in logic design and data processing", *Mathematics and Computers in Simulation*, 27, 1985, 431-439.

13. Sasao, T., "Representations of logic functions by using EXOR operators", in Sasao, T., Fujita, M., *Representations of Discrete Functions*, Kluwer Academic Publishers, Boston, 1996, 29-54.

14. Sasao, T., Fujita, M., (eds.), *Representations of Discrete Functions*, Kluwer Academic Publishers, Boston, 1996.

15. Shmerko, V.P., "Synthesis of arithmetical form of Boolean functions through the Fourier transform", *Automatics and Telemechanics*, No. 5, 1989, 134-142.

16. Stanković, M., Janković, D., Stanković, R.S., "Efficient algorithms for Haar spectrum calculation", *Scientific Review*, No. 21-22, 1996, 171-182.

17. Stanković, R.S., "Some remarks about spectral transform interpretation of MTB-DDs and EVBDDs", *Proc. ASP-DAC'95*, August 29-September 1, 1995, Makuhari Messe, Chiba, Japan, 385-390.

18. Stanković, R.S., *Spectral Transform Decision Diagrams in Simple Questions and Simple Answers*, Nauka, Belgrade, 1998.

19. Stankovic, R.S., Moraga, C., Astola, J.T., "From Fourier expansions to arithmetic-Haar expressions on quaternion groups", *Applicable Algebra in Engineering, Communication and Computing*, Vol. AAECC 12, 2001, 227-253.

20. Stanković, R.S., Sasao, T., Moraga, C., "Spectral transform decision diagrams" in Sasao, T., Fujita, M., *Representations of Discrete Functions*, Kluwer Academic Publishers, Boston, 1996, 55-92.

21. Stanković, R.S., Stanković, M., Janković, D., *Spectral Transforms in Switching Theory, Definitions and Calculations*, Nauka, Belgrade, 1998.

22. Yanushkevich, S.N., *Logic Differential Calculus in Multi-Valued Logic Design*, Techn. University of Szczecin Academic Publisher, Poland, 1998.

6

Gibbs Derivatives on Finite Groups

Differential operators are a very powerful tool for the mathematical modeling of natural phenomena. Usually the differentiation is with respect to time or to a spatial coordinate, modeled by the real line R. In this setting, using the differential operators, the direction, as well as the rate, of change of a quantity can be adequately described. Moreover, by forming linear differential equations with constant coefficients, we get a very convenient way of expressing the principle of superposition inherent in many natural phenomena. Linearity offers an easily tractable model, usually sufficiently good as a first approximation.

Fourier analysis, having linearity and the superposition principle in its essence, is another very efficient tool used for the same purposes.

It is known from classical analysis that there is a strong relationship between the Newton-Leibniz derivative f' of a function f on R and its Fourier transform, which can be expressed by

$$F'(w) = iwF(w), \tag{6.1}$$

where F' and F denote the Fourier transforms of f' and f, respectively.

Replacing the real group R by a locally compact Abelian, or a compact non-Abelian group extends classical Fourier analysis into abstract harmonic analysis. In this setting it has been natural to think about differentiation on groups in a way preserving as many as possible of the useful properties of Newton-Leibniz differentiation.

The Gibbs derivatives on groups [4], [8], [10], [20], [24], [33], [34], [40] form a class of differential operators extending the relation (6.1) to the functional spaces on other groups. In this more general context the role of the Euler functions $\exp(jtx)$ (the characters of the real group R) is taken over by the characters of locally compact

Abelian groups [4], [6], [8], [10], [18], [20], or by the unitary irreducible representations of compact non-Abelian groups [22]. In this book attention is focused on Gibbs derivatives on finite, not necessarily Abelian, groups. We consider in detail one of the matrix representations suggested in [11], which is suitable for the numerical evaluation of Gibbs derivatives of a given function. Using this matrix representation we present some FFT-like algorithms for calculation of the values of Gibbs derivatives on finite groups [29].

For a review of Gibbs differentiation see [30] and for some particular examples the bibliography [12] given in [3]. Some very recent results are reviewed in [30] and [38], [39].

6.1 DEFINITION AND PROPERTIES OF GIBBS DERIVATIVES ON FINITE NON-ABELIAN GROUPS

As is noted above, Gibbs differential operators on Abelian groups are defined as linear operators having the group characters as their eigenfunctions, see for example [10]. Since the group characters are the kernels of Fourier transforms on locally compact Abelian groups, it is very convenient to characterize the Gibbs derivatives by Fourier coefficients. Moreover, the strong relationship between the Gibbs derivatives and Fourier coefficients is somewhere used as the starting point for introduction of Gibbs derivatives on some particular groups, see for example [18]. By using the same approach the Gibbs derivatives on finite non-Abelian groups are defined in terms of Fourier coefficients as follows [22].

Definition 6.1 *Gibbs derivative Df of a function $f \in P(G)$ whose Fourier transform is S_f is defined by*

$$(Df)(x) = \sum_{w=0}^{K-1} wTr(S_f(w)R_w(x)). \tag{6.2}$$

As is noted in [22] this definition is unique only by virtue of the fixed order adopted for the elements of Γ. If a different notation was adopted, then (6.2), though unchanged in appearance, would define a distinct differentiator. This phenomenon is already present in the definition of the dyadic Gibbs derivative [7] which depends upon the order assumed for the Walsh functions. The same applies to all other Gibbs derivatives on various groups.

In what follows the Gibbs derivatives will be denoted by Df or, alternatively, by $f^{(1)}$.

An interpretation of the Gibbs derivative on finite non-Abelian groups can be given by following the approach used in [17] for the finite dyadic Gibbs derivative and later in [31] for the Gibbs derivatives on finite Abelian groups.

Define the partial sum $f_p(x)$, $p \le K$ by

$$f_p(x) = \sum_{w=0}^{p-1} Tr(\mathbf{S}_f(w)\mathbf{R}_w(x)). \tag{6.3}$$

Define also the Fejèr sum as

$$\sigma_q(x) = q^{-1} \sum_{p=1}^{q} f_p(x). \tag{6.4}$$

Substituting (6.3) into (6.4) we have, after a simple calculation,

$$f(x) - \sigma_K(x) = K^{-1} \sum_{w=0}^{K-1} wTr(\mathbf{S}_f(w)\mathbf{R}_w(x)).$$

The left member of this equality is the error in the approximation of f by its Fejèr sum $\sigma_K(x)$. Thus, the Gibbs derivative on a finite non-Abelian group G can be interpreted as that error multiplied by K.

The chief properties of Gibbs derivatives are analogs to the corresponding properties of the classical Newton-Leibniz derivative, and they are given by the following theorem.

Theorem 6.1 *If $f \in P(G)$, then*

1. $D(\alpha_1 f_1 + \alpha_2 f_2) = \alpha_1 D f_1 + \alpha_2 D f_2, \quad \alpha_1, \alpha_2 \in P, f_1, f_2 \in P(G).$

2. $Df = 0 \in P$ iff f is a constant function.

3. *If the Fourier transform of f is S_f, then that of $f^{(1)}$ is given by $S_{f^{(1)}}(w) = wS_f, \quad w = 0, \dots, K - 1.$*

   This property can be interpreted as the fact that the set $\{R_w^{(i,j)}(x)\}$ is the set of eigenfunctions of the Gibbs derivative, i.e.,

   $$DR_w^{(i,j)}(x) = wR_w^{(i,j)}(x).$$

   From that, thanks to the linearity of Gibbs derivatives,

   $$DTrR_w(x) = wTrR_w(x).$$

4. *From the property 3, it easily follows that*

   $$D(f_1 * f_2) = (Df_1) * f_2 = f_1 * (Df_2), \quad f_1, f_2 \in P(G),$$

   where $*$ denotes the convolution on G.

5. *The Gibbs derivative commutes with the translation (shift) operator T on G defined by $(T^\tau f)(x) = f(\tau \circ x^{-1})$, i.e.,*

   $$D(T^\tau f) = T^\tau (Df), \quad \text{for each } \tau \in G.$$

6. *It is known that the Gibbs differential operators do not obey the product rule. The same applies to the Gibbs derivatives on finite non-Abelian groups, i.e., it is false that for each f_1 and f_2*

   $$D(f_1 f_2) = f_1(Df_2) + (Df_1)f_2.$$

The Gibbs derivatives can be extended to an arbitrary complex order k by way of the definition of the delta function:

$$\delta(x) = g^{-1} \sum_{w=0}^{K-1} r_w Tr R_w(x).$$

The δ-function thus defined has the property

$$\delta(x) = \begin{cases} 1, & x = 0, \\ 0, & x \neq 0. \end{cases}$$

The Gibbs derivative of order k of the δ-function is obtained by a direct application of Property 3

$$\delta^{(k)}(x) = g^{-1} \sum_{w=0}^{K-1} w^k r_w Tr R_w(x).$$

By using Property 4 of Theorem 2.2:

$$(D^k f)(x) = ((D^k \delta) * f)(x) = \sum_{w=0}^{K-1} w^k Tr(\mathbf{S}_f(w) \mathbf{R}_w(x)). \tag{6.5}$$

6.2 GIBBS ANTI-DERIVATIVE

In this section we consider the determination of the values of a function from the values of its Gibbs derivative.

It is obvious from (6.5) that the Gibbs derivative can be considered as a convolution operator on $P(G)$. More precisely, if we introduce a function W_k defined by its Fourier coefficients as

$$S_{W_k}(w) = \begin{cases} 0, & w = 0, \\ r_w g^{-1} w^k \mathbf{I}_{r_w}, & w = 1, \ldots, K-1, \end{cases}$$

where $\mathbf{I}_{r_w}$ is the $(r_w \times r_w)$ identity matrix, then from (6.5) the Gibbs derivative of order k of a function $f \in P(G)$ is given by

$$(Df)(w) = (W_k * f)(x).$$

From here we immediately deduce the concept of the Gibbs anti-derivative. Introduce a function W_{-k} defined in the transform domain by

$$S_{W_{-k}}(w) = \begin{cases} 1, & w = 0 \\ r_w w^{-k} \mathbf{I}_{r_w} & w = 1, \ldots, K-1. \end{cases}$$

After the inverse Fourier transform

$$W_{-k}(x) = 1 + \sum_{w=1}^{K-1} w^{-k} r_w Tr(\mathbf{R}_w(x)).$$

Functions of this kind for the particular case of the dyadic group were apparently first investigated in [37] in the Walsh-Fourier multiplier theory. Such functions were later used for the dyadic derivatives in [4] for the same purposes as those considered here. To be consistent with these particular definitions we omitted the factor g^{-1} the appearance of which could be expected from the convolution theorem.

By using the function W_{-k} we introduce an inverse operator called the Gibbs anti-derivative on finite non-Abelian groups [23]

Definition 6.2 *For a function $f \in P(G)$ the Gibbs anti-derivative of order k, denoted by I^k, is defined by*

$$(I^k f)(x) = (W_k * f)(x).$$

The Gibbs anti-derivative can be considered as a Fourier multiplier operator, thus having all properties characteristic for these operators. Therefore, there is no need for any particular consideration of these properties here.

Having the concept of Gibbs anti-derivative, we can deduce a theorem which shows how to determine the values of a function f from the values of its Gibbs derivative of order k.

Theorem 6.2 *Let $f \in P(G)$ be such that $S_f(0) = 0$. Then,*

$$f(x) = g^{-1} I^k (D^k f)(x),$$

or, equivalently,

$$f(x) = g^{-1} D^k (I^k f)(x).$$

Here the factor g^{-1} appears at the right hand side of the above equations since it was omitted in the definition of $S_{W^{-k}}$.

Note that Theorem 6.2 can be regarded as a kind of counterpart of the so-called fundamental theorem for dyadic analysis due to Butzer and Wagner [4]. Moreover, as is noted in [2], see also [21], theorems of this kind are a kind of counterpart of the fundamental theorem of the Newton-Leibniz calculus in abstract harmonic analysis.

6.3   PARTIAL GIBBS DERIVATIVES

As we noted above, a given function $f \in P(G)$, G-decomposable group, can be viewed as a function of several variables $f(x_1, \ldots, x_n)$, $x_i \in G_i$. Therefore, partial Gibbs derivative of f with respect to the variable x_i can be defined [26].

From Definition 6.1 and some well-known properties of unitary irreducible representations

$$(Df)(x) = \sum_{w=0}^{K-1} wTr(r_w g^{-1} \sum_{u=0}^{g-1} f(u)\mathbf{R}_w(u^{-1} \circ x)).$$

From the invariance under translation of the Haar integral

$$\sum_{y \in G} g(y) = \sum_{y \in G} g(z \circ y), \forall z \in G, g \in P(G),$$

it follows

$$(Df)(x) = g^{-1} \sum_{u=0}^{g-1} f(u \circ x) \sum_{w=0}^{K-1} w r_w Tr(\mathbf{R}_w(u^{-1})),$$

which suggests the following definition.

Definition 6.3 *Partial Gibbs derivative* $(\Delta_i f)(x)$ *at a point*
$x = (x_1, \ldots, x_{i-1}, x_i, x_{i+1}, \ldots, x_n) \in G$ *with respect to the i-th variable* x_i *of a function* $f \in P(G)$ *is defined as the Gibbs derivative* $(Df_i)(x_i)$, *at* x_i, *of the function* $f_i(y) = f(x_1, \ldots, x_{i-1}, y, x_{i+1}, \ldots, x_n))$. *Thus,*

$$(\Delta_i f)(x) = (Df_i)(x_i) =$$

$$g^{-1} \sum_{u_i}^{g_i - 1} f(x_1, \ldots, x_{i-1}, u_i \overset{i}{\circ}, x_{i+1}, \ldots, x_n) \sum_{w=0}^{K_i - 1} w r_w^i Tr(\mathbf{R}_w^i(u_i^{-1})),$$

where g_i *is the order of* G_i, K_i *denotes the number of nonequivalent unitary irreducible representations of* G_i, *and* r_w^i *is the dimension of the representation* $\mathbf{R}_w^i$ *of* G_i.

Actually, the partial Gibbs differentiator Δ_i thus defined is the restriction on G_i of the Gibbs differentiator on G. It follows that the partial Gibbs derivatives have properties corresponding to those of the Gibbs derivative.

Theorem 6.3 *Let* $f \in P(G)$, *Then,*

1. $\Delta_i f = 0$ *iff* f *is a constant on* G_i, *i.e., iff* f *has the same value for each* $x_I \in G_i$. *Moreover,* $\Delta_i c = 0$ *for any constant* $c \in P(G)$.

2. $\Delta_i(c_1 f_1 + c_2 f_2) = c_1 \Delta_i f_1 + c_2 \Delta_i f_2$, $c_1, c_2 \in P$, $f_1, f_2 \in P(G)$.

3. *If the Fourier transform of* $f \in P(G)$ *is* S_f, *then that of* $\Delta_i f$ *is given by*

$$S_{\Delta_i f}(w) = B_i(w) S_f(w), \quad w = 0, \ldots, K = 1,$$

   where $B_i(w)$ *is given by the vector of order g of the form*

$$B_i(w) = [0, 1, \ldots, K_i - 1, 0, 1, \ldots, K_i - 1, \ldots, 0, 1, \ldots, K_i - 1]^T.$$

4. $\Delta_i(f * g) = \Delta_i f * g = f * \Delta_i g$.

6.4 GIBBS DIFFERENTIAL EQUATIONS

Relation (6.5) introduces the Gibbs derivative of an arbitrary complex order. The Gibbs derivative of a positive integer order n can be defined recursively by $D^{n+1}f = D(D^n f)$, $n = 1, 2, \ldots$. This permits linear equations with constant coefficients in terms of Gibbs derivatives to be defined and solved. These equations can be considered as a particular case of the generalized linear equations studied in [16].

Definition 6.4 *A linear Gibbs discrete differential equation with constant coefficients is an equation of the form*

$$\sum_{k=0}^{n} a_k y^{(k)} = \sum_{k=0}^{m} b_k f^{(k)},\tag{6.6}$$

where a_k, b_k are real numbers, $f \in P(G)$ and y is the required solution.

As in the case of ordinary differential equations we get the general solution, y, of the equation (6.6) as the sum of the solution y_{zi} of the homogeneous equation and the partial solution y_{zs} of the inhomogeneous equation, i.e.,

$$y = y_{zi} + y_{zs}.\tag{6.7}$$

In order to find y_{zi} one looks for roots of the characteristic equation of (6.6)

$$\sum_{k=0}^{n} a_k z^k = 0.$$

Now, we have the following theorem.

Theorem 6.4 *If the roots $\{z_i\}$, $i = 0, \ldots, n$ of the characteristic equation are distinct and belong to the set $\{0, \ldots, K-1\}$, then the homogeneous solution of (6.6) is*

$$y_{zi}(x) = \sum_{i=0}^{n} \sum_{j,k=1}^{i} c_{jk}^{z_i} R_{z_i}^{(j,k)}(x),$$

where the constants $c_{jk}^{z_i}$ depend on the boundary conditions.

There are some important differences encountered in solving linear Gibbs discrete differential equations with constant coefficients compared to solving of ordinary differential equations. A homogeneous equation of order n does not always have n linearly independent solutions. The following statements are, in a way, often taken for granted, however, we could not find a proof for these statements, anywhere.

If t roots of the characteristic equation are repetitions of the other roots, then the number of linearly independent solutions of a linear Gibbs discrete differential equation of order k is $\sum_{i=0}^{k-t} r_{z_i}^2$, provided that each root of the characteristic equation is in the set $\{0, \ldots, K-1\}$.

If s of the roots are not in this set, then the number of linearly independent solutions of the given equation is $\sum_{i=0}^{k-s-t} r_{z_i}^2$. This is not any peculiarity of the case considered here. A corresponding statement for so-called logical differential equations with Gibbs derivatives on the dyadic group is given in [9]. Moreover, it seems that an analogous statement is valid more generally, as it is noted without proof in [16].

To get the particular solution of (6.6) we apply the Fourier transform on both sides of (6.6), and with property 3 of Theorem 6.1 we obtain:

$$\sum_{k=0}^{n} a_k w^k S_y(w) \sum_{k=0}^{m} b_k w^k S_f(w). \tag{6.8}$$

From there, providing that equation (6.8) is compatible, that is, $S_f(0) = 0$ for all $w \in \{z_i\}, i = 0, \ldots, n$, we have

$$S_y(w) = \frac{P}{Q} S_f(w),$$

where

$$P = \sum_{k=0}^{m} b_k w^k, \quad Q = \sum_{k=0}^{m} a_k w^k.$$

By introducing the notation

$$H(w) = r_w g^{-1} \frac{P}{Q}, \tag{6.9}$$

we have

$$S_y(w) = r_w^{-1} H(w) S_f(w). \tag{6.10}$$

From (6.10) by using the convolution property, the inverse Fourier transform produces the particular solution

$$y_{zs}(x) = \sum_{u=0}^{g-1} h(u) f(xu^{-1}). \tag{6.11}$$

It follows that (6.6) has a general solution of the form

$$y(x) = \sum_{i=0}^{n} \sum_{j,k=1}^{m} c_{jk}^{z_i} R_{z_i}^{(j,k)}(x) + \sum_{u=0}^{g-1} h(u) f(xu^{-1}).$$

6.5   MATRIX INTERPRETATION OF GIBBS DERIVATIVES

In this section we will consider a matrix representation of Gibbs differential operators suitable for their numerical evaluation. One of the main properties characterizing Gibbs derivatives is given by

$$\mathbf{S}_{Df}(w) = w \mathbf{S}_f(w), \quad w \in \{0, 1, \ldots, K-1\}, \tag{6.12}$$

where $\mathbf{S}_f(w)$ are the Fourier coefficients of a function $f \in P(G)$, while $\mathbf{S}_{Df}(w)$ denotes the Fourier coefficients of its Gibbs derivative $D_g f$. Moreover, a wish to have a differential operator satisfying this property motivated the introduction of the class of differential operators considered here. In this setting, the relation (6.12) is used by some authors as a starting point for defining certain Gibbs derivatives on groups in terms of formal Fourier series (see, for example, [18],[24], [25], [40], [33], [34]). Using this approach, the Gibbs derivative on a finite group can be defined in matrix notation as follows.

Definition 6.5 *The Gibbs derivative* $\mathbf{D}_g$ *on a finite, not necessarily Abelian, group* G *of order* g *is defined [15], [29] as*

$$\mathbf{D}_g = g^{-1}[\mathbf{R}] \circ \mathbf{G} \odot [\mathbf{R}]^{-1},$$

where $[\mathbf{R}]$ *is the matrix of unitary irreducible representations of* G *over* P, *i.e.,* $[\mathbf{R}] = [\mathbf{a}_{ij}]$ *with* $\mathbf{a}_{ij} = \mathbf{R}_j(i)$, $j \in \{0, 1, \ldots, g-1\}$, $j \in \{0, 1, \ldots, K-1\}$, $\mathbf{G}$ *is a diagonal* $(K \times K)$ *matrix given by* $\mathbf{G} = diag(0, 1, \ldots, K-1)$, *and* $[\mathbf{R}]^{-1} = [\mathbf{b}_{sq}]$ *with* $\mathbf{b}_{sq} = r_s \mathbf{R}_s^{-1}(q)$, $s \in \{0, 1, \ldots, K-1\}$, $q \in \{0, 1, \ldots, g-1\}$.

For a function $f(x) = f(x_1, \ldots, x_n) \in P$ we define the partial Gibbs derivative with respect to the variable $x_i \in G$ as a restriction on G_i of the previously introduced Gibbs derivative on G.

Definition 6.6 *Let* G *be representable in the form (2.4). The partial Gibbs derivative* Δ_i *with respect to the variable* x_i *is defined as:*

$$\Delta_i = \bigotimes_{i=1}^{n} \mathbf{A}_j,$$

with

$$\mathbf{A}_j = \begin{cases} g_j \mathbf{D}_{g_j}, & j = i, \\ \mathbf{I}_{(g_j \times g_j)}, & j \neq i, \end{cases}$$

where $\mathbf{I}_{(g_j \times g_j)}$ *is a* $(g_j \times g_j)$ *identity matrix, and* $\otimes$ *denotes the Kronecker product.*

Note that, using the representation (2.13) for the first K non-negative integers $\{0, 1, \ldots, K-1\}$, the matrix $\mathbf{G}$ can be expressed as:

$$\mathbf{G} = \sum_{i=1}^{n} \mathbf{w}_i \mathbf{Q}_i,$$

where,

$$\mathbf{Q}_i = \bigotimes_{i=1}^{n} \mathbf{Z}_j^i,$$

with

$$\mathbf{Z}_j^i = \begin{cases} \mathbf{G}_j, & i = j, \\ \mathbf{I}_{(K_i \times K_i)}, & i \neq j, \end{cases}$$

where $\mathbf{G}_j$ is a diagonal $(K_j \times K_j)$ matrix given by $\mathbf{G}_j = diag(0, b_j, b_j, \ldots, b_j)$ with b_j defined by (2.13).

Recall that the matrix $[\mathbf{R}]$ is the matrix of unitary irreducible representations of G over P. Since G is representable in the form (2.4), the matrix $[\mathbf{R}]$ can be generated as the Kronecker product of $(K_i \times g_i)$ matrices $[\mathbf{R}_i]$ of unitary irreducible representations of subgroups G_i, $i = 1, \ldots, n$, i.e.,

$$[\mathbf{R}] = \bigotimes_{i=1}^{n} [\mathbf{R}_i].$$

Thanks to the well-known properties of the Kronecker product, the same applies to the matrix $[\mathbf{R}]^{-1}$, i.e, for this matrix holds

$$[\mathbf{R}]^{-1} = \bigotimes_{i=1}^{n} [\mathbf{R}_i]^{-1}.$$

By using the representations introduced above for the matrices $[\mathbf{R}]$, $[\mathbf{R}]^{-1}$ and $\mathbf{G}$, the matrix $\mathbf{D}_g$ of the Gibbs derivative can be rewritten as

$$\mathbf{D}_g = g^{-1} \left[\bigotimes_{i=1}^{n} [\mathbf{R}_i] \right] \circ \left[\sum_{i=1}^{n} \mathbf{w}_i \mathbf{Q}_i \right] \odot \left[\bigotimes_{i=1}^{n} [\mathbf{R}_i]^{-1} \right].$$

After a short calculation, by using well-known properties of the Kronecker product we prove the following.

Proposition 6.1 *The matrix $\mathbf{D}_g$ representing the Gibbs derivative on a finite group G of order g can be expressed in terms of partial Gibbs derivatives as*

$$\mathbf{D}_g = \sum_{i=1}^{n} b_i \Delta_i, \tag{6.13}$$

where the coefficients b_i are defined by (2.8).

6.6 FAST ALGORITHMS FOR CALCULATION OF GIBBS DERIVATIVES ON FINITE GROUPS

In this section we will disclose fast algorithms for computation of Gibbs derivatives on finite groups. As is noted in [29], the application of Definition 6.5 leads to an algorithm for the computation of Gibbs derivatives of which the complexity is obviously approximately equal to the complexity of calculation of one direct and one inverse

Fourier transform. The advantage is that the application of fast Fourier transform on groups is immediately possible without any considerable modification. However, from the computational point of view a more efficient algorithm for the computation of Gibbs derivatives on finite groups can be disclosed defining the Gibbs derivative in terms of partial Gibbs derivatives, that is, starting from the relation (6.13). Moreover, the algorithm thus obtained is quite suitable for a parallel implementation. The idea comes from the following facts.

As is noted in Section 3.1, the definition of the fast Fourier transform (FFT) on group G is based upon the factorization of G into the equivalence classes relative to some subgroups of G. On the other hand, the i-th partial Gibbs derivative on a group G representable in the form (2.4) is defined as the restriction of Gibbs differentiation on G to the differentiation on G_i. Therefore, it is natural to search for a fast algorithm for the computation of Gibbs derivatives through the partial Gibbs derivatives.

Note that the i-th partial Gibbs derivative is defined (Definition 6.6) by a relation of the form (3.1). Therefore, comparing the matrices Δ_i, $i \in \{1, \ldots, n\}$, for a given group G with the matrices $[\mathbf{C}^{n-k}]$ appearing in the factorization of the Fourier transformation matrix, we infer a strong similarity. It follows that the algorithm for the computation of i-th partial Gibbs derivative will be similar to the i-th step of the FFT, and, hence, may be described by a flow-graph similar to that describing the i-th step of the FFT. Naturally, the similarity is less strong in the case of non-Abelian groups than in the case of Abelian groups, for the dual object Γ of a non-Abelian group G does not have the structure of a group isomorphic with G as is the case for Abelian groups. More precisely, the cardinality of Γ is not equal to the order of G, and the consequence is that in the case of non-Abelian groups the number of input nodes in the flow-graph is different for each step of the FFT, as we noted above, while all partial Gibbs derivatives are by definition applicable to the vector $\mathbf{f}$ whose order is g. Nevertheless, the similarity in the overall structure of the corresponding flow-graphs is retained and can be efficiently used for the disclosure of the fast algorithms for the computation of partial Gibbs derivatives.

As we noted above, the flow-graph for the computation of the k-th partial Gibbs derivative of a function f on a finite group G of order g consists of g input and g output nodes of which some are connected by branches. It is determined by the overall structure of the matrix Δ_k which nodes will be mutually connected, as in the case of the FFT. More precisely, the output node j will be connected with the input node i iff the element d_{ij} of Δ_k describing the k-th partial Gibbs derivative is not equal to zero. As in the case of the FFT, a weighting coefficient is associated to each branch. However, all weights in the fast algorithm for the computation of Gibbs derivatives are numbers belonging to P even for non-Abelian groups, which is another considerable difference relative to FFT for this case. Denoting by $k(i, j)$ the branch connecting the output node i with the input node j, the weight $w^k(i, j)$ associated with this branch is given by $w(i, j) = d_{ij}$, where d_{ij} is the (i, j)-th element of Δ_k.

Having the fast algorithms for the computation of partial Gibbs derivatives, one obtains the fast algorithm for the computation of the Gibbs derivative of a function $f \in P(G)$ according to (6.13) simply by adding the output nodes of the flow-graphs for the calculation of partial Gibbs derivatives multiplied by the weight coefficients b_i defined by (2.13).

Now, we give a brief analysis of the complexity of the algorithm described above.

The number of calculations is usually employed as a first approximation to the complexity of an algorithm.

It is obvious from Definition 6.6 that the number of calculations required to calculate $\Delta_i f$ is equal to gg_i, since there are at most g_i non-zero elements in each row of the matrix Δ_i. According to (6.13), the number of operations needed to calculate the Gibbs derivative $D_g f$ is equal to $g(\sum_{i=1}^{n} g_i)$ followed by $(n-1)g$ multiplications with the weighting factors b_i and ng additions. Recall that the number of calculations in an FFT to which our algorithm can be compared is $g(\sum_{i=1}^{n} g_i)$.

However, to confirm the efficiency of the algorithm proposed it is important to give at least a rough estimate of its overall time complexity.

It is obvious from Definition 6.6 that, unlike the FFT, which is a sequential algorithm in its essence for the input to one stage is the output from the preceding stage, the fast algorithm for the calculation of Gibbs derivatives (FGD) is quite suitable for a parallel implementation, since the calculation of the partial Gibbs derivatives can be carried out simultaneously. This parallelism is over and above the parallelism in each step of the calculation of each Δ_I as in the corresponding step of the FFT. It follows that the time complexity of the FGD does not depend on the number of subgroups of G as is the case with the FFT. The FGD is considerably faster than the corresponding FFT, since it can always be implemented in only two steps, compared with the n steps required in the FFT. Of course, the price is the number of processors operating in parallel and the memory storage requirements, which in this case should certainly be greater for the FGD than for the FFT.

A more accurate analysis of the complexity of the FGD is certainly needed, but to be correct it may only be done after rather precise specification of the facilities used for the implementation.

As is usually the case in the study of problems like that considered here, the procedure for the numerical calculation is best explained by some examples. We shall therefore consider two examples, the first for Abelian groups and the second for non-Abelian groups.

Example 6.1 *Let* $G = Z_9 = (\{0, 1, 2, 3, 4, 5, 6, 7, 8\}, \circ)$ *be the group of non-negative integers less than 9 with componentwise addition modulo 3 of 3-adic expansions of group elements as the group operation. For convenience the group operation is shown in Table 6.1. The group representations of Z_9 over the complex field are the Vilenkin-Chrestenson functions shown in a matrix form in Table 6.2. Note that the group Z_9 can be considered as the product $Z_9 = Z_3 \times Z_3$ where $Z_3 = (\{0, 1, 2\}, 3)$ is the group of non-negat66ive integers less than 3 with addition modulo 3 as group operation. Therefore, any complex-valued function f on Z_9 can be considered as a two-variable function $f(x_1, x_2)$, $x_1, x_2 \in Z_3$.*

The matrices Δ_1^9 and Δ_2^9 of the partial Gibbs derivatives relative to the variables x_1 and x_2, respectively, and the matrix $\mathbf{D}_9$ of the Gibbs derivative on Z_9 calculated according to (6.13) as $\mathbf{D}_9 = 3\Delta_1^9 + \Delta_1^9$ are shown in Figure 6.1 a, b, c, respectively.

Table 6.1 Group operation of Z_9.

∘	0	1	2	3	4	5	6	7	8
1	1	2	0	4	5	3	7	8	6
2	2	0	1	5	6	7	8	6	7
3	3	4	5	6	7	8	0	1	2
4	4	5	3	7	8	6	1	2	0
5	5	3	4	8	6	7	2	0	1
6	6	7	8	0	1	2	3	4	5
7	7	8	6	1	2	0	4	5	3
8	8	6	7	2	0	1	5	3	4

Table 6.2 The group representations of Z_9 over C.

$$
\begin{bmatrix}
1 & 1 & 1 & 1 & 1 & 1 & 1 & 1 & 1 \\
1 & e_1 & e_2 & 1 & e_1 & e_2 & 1 & e_1 & e_2 \\
1 & e_2 & e_1 & 1 & e_2 & e_1 & 1 & e_2 & e_1 \\
1 & 1 & 1 & e_1 & e_1 & e_1 & e_2 & e_2 & e_2 \\
1 & e_1 & e_2 & e_1 & e_2 & 1 & e_2 & 1 & e_1 \\
1 & e_2 & e_1 & e_1 & 1 & e_2 & e_2 & e_1 & 1 \\
1 & 1 & 1 & e_2 & e_2 & e_2 & e_1 & e_1 & e_1 \\
1 & e_1 & e_2 & e_2 & 1 & e_1 & e_1 & e_2 & 1 \\
1 & e_2 & e_1 & e_2 & e_1 & 1 & e_1 & 1 & e_2
\end{bmatrix},
$$

$$
e_1 = -\tfrac{1}{2}(1 - i\sqrt{3}), \quad e_2 - \tfrac{1}{2}(1 + i\sqrt{3})
$$

The flow-graph of the fast algorithm for the computation of the Gibbs derivative $\mathbf{D}_9$ of a complex-valued function f on Z_9 given by its truth vector $\mathbf{f} = [f(0), \dots, f(8)]^T$ is shown in Figure 6.2.

Now, let us consider as the second example the calculation of the Gibbs derivatives of functions defined on the non-Abelian group of binary matrices described in [15].

Example 6.2 *Let G be the multiplication group of the twelve (3×3) matrices $\mathbf{t} = [t_{ij}]$, $i, j = 0, 1, 2$, over the complex field represented in Table 6.4. For convenience the group operation is explicitly shown in Table 6.3. Note that G is isomorphic to the direct product of the cyclic group $C_2 = (\{0, 1\}, \circ)$ of order 2 with generating*

$$\Delta_1^9 = \begin{bmatrix} 1 & 0 & 0 & a & 0 & 0 & b & 0 & 0 \\ 0 & 1 & 0 & 0 & a & 0 & 0 & b & 0 \\ 0 & 0 & 1 & 0 & 0 & a & 0 & 0 & b \\ b & 0 & 0 & 1 & 0 & 0 & a & 0 & 0 \\ 0 & b & 0 & 0 & 1 & 0 & 0 & a & 0 \\ 0 & 0 & b & 0 & 0 & 1 & 0 & 0 & a \\ a & 0 & 0 & b & 0 & 0 & 1 & 0 & 0 \\ 0 & a & 0 & 0 & b & 0 & 0 & 1 & 0 \\ 0 & 0 & a & 0 & 0 & b & 0 & 0 & 1 \end{bmatrix},$$

a.

$$\Delta_2^9 = \begin{bmatrix} 1 & a & b & 0 & 0 & 0 & 0 & 0 & 0 \\ b & 1 & a & 0 & 0 & 0 & 0 & 0 & 0 \\ a & b & 1 & 0 & 0 & 0 & 0 & 0 & 0 \\ 0 & 0 & 0 & 1 & a & b & 0 & 0 & 0 \\ 0 & 0 & 0 & b & 1 & a & 0 & 0 & 0 \\ 0 & 0 & 0 & a & b & 1 & 0 & 0 & 0 \\ 0 & 0 & 0 & 0 & 0 & 0 & 1 & a & b \\ 0 & 0 & 0 & 0 & 0 & 0 & b & 1 & a \\ 0 & 0 & 0 & 0 & 0 & 0 & a & b & 1 \end{bmatrix},$$

b.

$$\mathbf{D}_9 = \begin{bmatrix} 4 & a & b & 3a & 0 & 0 & 3b & 0 & 0 \\ b & 4 & a & 0 & 3a & 0 & 0 & 3b & 0 \\ a & b & 4 & 0 & 0 & 3a & 0 & 0 & 3b \\ 3b & 0 & 0 & 4 & a & b & 3a & 0 & 0 \\ 0 & 3b & 0 & b & 4 & a & 0 & 3a & 0 \\ 0 & 0 & 3b & a & b & 4 & 0 & 0 & 3a \\ 3a & 0 & 0 & 3b & 0 & 0 & 4 & a & b \\ 0 & 3a & 0 & 0 & 3b & 0 & b & 4 & a \\ 0 & 0 & 3a & 0 & 0 & 3b & a & b & 4 \end{bmatrix},$$

c.

$$a = \tfrac{1}{3}(e_1 - 1), \quad b = \tfrac{1}{3}(e_2 - 1)$$

Fig. 6.1 *a.* The partial Gibbs derivative Δ_1^9 on Z_9, *b.* The partial Gibbs derivative Δ_2^9 on Z_9, *c.* The Gibbs derivative $\mathbf{D}_9$ on Z_9.

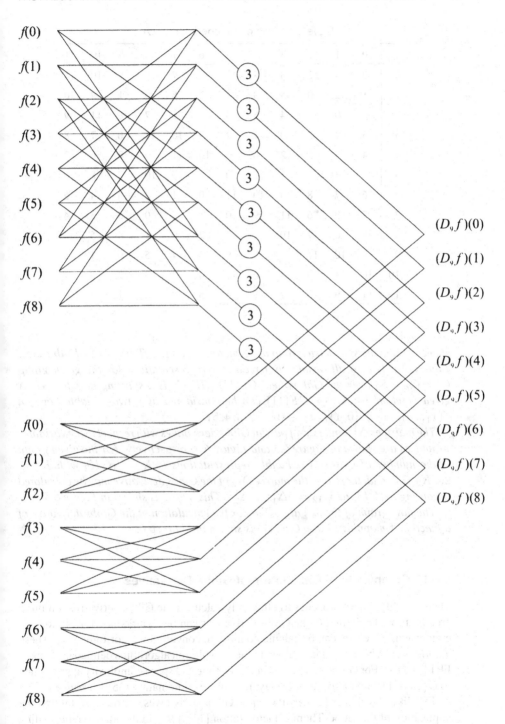

Fig. 6.2 The flow-graph of the fast algorithm for calculation of the Gibbs derivative D_9 on Z_9.

Table 6.3 The group operation of G_{12}.

∘	0	1	2	3	4	5	6	7	8	9	10	11
0	0	1	2	3	4	5	6	7	8	9	10	11
1	1	2	0	5	3	4	7	8	6	11	9	10
2	2	0	1	4	5	3	8	6	7	10	11	9
3	3	4	5	0	1	2	9	10	11	6	7	8
4	4	5	3	2	0	1	10	11	9	8	6	7
5	5	3	4	1	2	0	11	9	10	7	8	6
6	6	7	8	9	10	11	0	1	2	3	4	5
7	7	8	6	11	9	10	1	2	0	5	3	4
8	8	6	7	10	11	9	2	0	1	4	5	3
9	9	10	11	6	7	8	3	4	5	0	1	2
10	10	11	9	8	6	7	4	5	3	2	0	1
11	11	9	10	7	8	6	5	3	4	1	2	0

element 1 and the symmetric group of permutations S_3 (see Table 2.1 and Table 2.5). Table 6.4 lists also all absolutely irreducible representations for the given group $G = C_2 \times S_3$ over the Galois field GF(11) (GF(11) is a splitting field for G). A given function $f : G \rightarrow GF(11)$ can be considered as a two-variable function $f(x_1, x_2)$, $x_1 \in \{0, 1\}$, $x_2 \in \{0, 1, 2, 3, 4, 5\}$.

The matrices Δ_1^{12} and Δ_2^{12} of partial Gibbs derivatives with respect to the variables x_1 and x_2 are shown in Figure 6.3 and Figure 6.4, respectively. For the given group G the number of unitary irreducible representations is $K = K_1 K_2$ with $K_1 = 2$, $K_2 = 3$, and, therefore, the matrix $\mathbf{D}_{12}$ of the Gibbs derivative may be calculated according to (6.13) as $\mathbf{D}_{12} = 3\Delta_1^{12} + \Delta_2^{12}$. This matrix is shown in Figure 6.5.

The flow-graph of the fast algorithm for the calculation of the Gibbs derivative of a function f mapping G into $GF(11)$ is shown in Figure 6.6.

6.6.1 Complexity of Calculation of Gibbs Derivatives

There are different approaches to efficiently calculate the Gibbs derivatives on finite groups. From Definition 6, the Gibbs derivatives can be regarded as convolution operators and, therefore, can be calculated through convolution algorithms [19]. These algorithms can be derived in analogy to the convolution algorithms defined in terms of FFT [1], [19]. For Gibbs derivatives in $P(G)$, time and space complexity approximate to $O(2n + 1)$, and $O(g)$, respectively, if in-place computation [35] is assumed.

FFT-like algorithms for calculation of Gibbs derivatives are derived through the application of the Good-Thomas factorization [13], [36], to the matrix representing the Gibbs derivative on a given group G [29]. Unlike the algorithms for calculation

Table 6.4 The representations of G_{12} over $GF(11)$.

x	t	R_0	R_1	R_2	R_3	R_4	R_5
0	$\begin{bmatrix} 1 & 0 & 0 \\ 0 & 1 & 0 \\ 0 & 0 & 1 \end{bmatrix}$	1	1	$\begin{bmatrix} 1 & 0 \\ 0 & 1 \end{bmatrix}$	1	1	$\begin{bmatrix} 1 & 0 \\ 0 & 1 \end{bmatrix}$
1	$\begin{bmatrix} 1 & 0 & 0 \\ 0 & -1 & 1 \\ 0 & -1 & 0 \end{bmatrix}$	1	1	$\begin{bmatrix} 5 & 8 \\ 3 & 5 \end{bmatrix}$	1	1	$\begin{bmatrix} 5 & 8 \\ 3 & 5 \end{bmatrix}$
2	$\begin{bmatrix} 1 & 0 & 0 \\ 0 & 0 & -1 \\ 0 & 1 & -1 \end{bmatrix}$	1	1	$\begin{bmatrix} 5 & 3 \\ 8 & 5 \end{bmatrix}$	1	1	$\begin{bmatrix} 5 & 3 \\ 8 & 5 \end{bmatrix}$
3	$\begin{bmatrix} 1 & 0 & 0 \\ 0 & -1 & 1 \\ 0 & 0 & 1 \end{bmatrix}$	1	10	$\begin{bmatrix} 1 & 0 \\ 0 & 10 \end{bmatrix}$	1	10	$\begin{bmatrix} 1 & 0 \\ 0 & 10 \end{bmatrix}$
4	$\begin{bmatrix} 1 & 0 & 0 \\ 0 & 0 & -1 \\ 0 & -1 & 0 \end{bmatrix}$	1	10	$\begin{bmatrix} 5 & 8 \\ 8 & 6 \end{bmatrix}$	1	10	$\begin{bmatrix} 5 & 8 \\ 8 & 6 \end{bmatrix}$
5	$\begin{bmatrix} 1 & 0 & 0 \\ 0 & 1 & 0 \\ 0 & 1 & -1 \end{bmatrix}$	1	10	$\begin{bmatrix} 5 & 3 \\ 3 & 6 \end{bmatrix}$	1	10	$\begin{bmatrix} 5 & 3 \\ 3 & 6 \end{bmatrix}$
6	$\begin{bmatrix} -1 & 0 & 0 \\ 0 & 1 & 0 \\ 0 & 0 & 1 \end{bmatrix}$	1	1	$\begin{bmatrix} 1 & 0 \\ 0 & 1 \end{bmatrix}$	10	10	$\begin{bmatrix} 10 & 0 \\ 0 & 10 \end{bmatrix}$
7	$\begin{bmatrix} -1 & 0 & 0 \\ 0 & -1 & 1 \\ 0 & -1 & 0 \end{bmatrix}$	1	1	$\begin{bmatrix} 5 & 8 \\ 3 & 5 \end{bmatrix}$	10	10	$\begin{bmatrix} 6 & 3 \\ 8 & 6 \end{bmatrix}$
8	$\begin{bmatrix} -1 & 0 & 0 \\ 0 & 0 & -1 \\ 0 & 1 & -1 \end{bmatrix}$	1	1	$\begin{bmatrix} 5 & 3 \\ 8 & 5 \end{bmatrix}$	10	10	$\begin{bmatrix} 6 & 8 \\ 3 & 6 \end{bmatrix}$
9	$\begin{bmatrix} -1 & 0 & 0 \\ 0 & -1 & 1 \\ 0 & 0 & 1 \end{bmatrix}$	1	10	$\begin{bmatrix} 1 & 0 \\ 0 & 10 \end{bmatrix}$	10	1	$\begin{bmatrix} 10 & 0 \\ 0 & 1 \end{bmatrix}$
10	$\begin{bmatrix} -1 & 0 & 0 \\ 0 & 0 & -1 \\ 0 & -1 & 0 \end{bmatrix}$	1	10	$\begin{bmatrix} 5 & 8 \\ 8 & 6 \end{bmatrix}$	10	1	$\begin{bmatrix} 6 & 3 \\ 3 & 5 \end{bmatrix}$
11	$\begin{bmatrix} 1 & 0 & 0 \\ 0 & 1 & 0 \\ 0 & 1 & -1 \end{bmatrix}$	1	10	$\begin{bmatrix} 5 & 3 \\ 3 & 6 \end{bmatrix}$	10	1	$\begin{bmatrix} 6 & 8 \\ 8 & 5 \end{bmatrix}$

$$\Delta_1^{12} = \begin{bmatrix} 6 & 0 & 0 & 0 & 0 & 0 & 5 & 0 & 0 & 0 & 0 & 0 \\ 0 & 6 & 0 & 0 & 0 & 0 & 0 & 5 & 0 & 0 & 0 & 0 \\ 0 & 0 & 6 & 0 & 0 & 0 & 0 & 0 & 5 & 0 & 0 & 0 \\ 0 & 0 & 0 & 6 & 0 & 0 & 0 & 0 & 0 & 5 & 0 & 0 \\ 0 & 0 & 0 & 0 & 6 & 0 & 0 & 0 & 0 & 0 & 5 & 0 \\ 0 & 0 & 0 & 0 & 0 & 6 & 0 & 0 & 0 & 0 & 0 & 5 \\ 5 & 0 & 0 & 0 & 0 & 0 & 6 & 0 & 0 & 0 & 0 & 0 \\ 0 & 5 & 0 & 0 & 0 & 0 & 0 & 6 & 0 & 0 & 0 & 0 \\ 0 & 0 & 5 & 0 & 0 & 0 & 0 & 0 & 6 & 0 & 0 & 0 \\ 0 & 0 & 0 & 5 & 0 & 0 & 0 & 0 & 0 & 6 & 0 & 0 \\ 0 & 0 & 0 & 0 & 5 & 0 & 0 & 0 & 0 & 0 & 6 & 0 \\ 0 & 0 & 0 & 0 & 0 & 5 & 0 & 0 & 0 & 0 & 0 & 6 \end{bmatrix}$$

Fig. 6.3 The partial Gibbs derivative Δ_1^{12} on G_{12}.

$$\Delta_2^{12} = \begin{bmatrix} 7 & 5 & 5 & 9 & 9 & 9 & 0 & 0 & 0 & 0 & 0 & 0 \\ 5 & 7 & 5 & 9 & 9 & 9 & 0 & 0 & 0 & 0 & 0 & 0 \\ 5 & 5 & 7 & 9 & 9 & 9 & 0 & 0 & 0 & 0 & 0 & 0 \\ 9 & 9 & 9 & 7 & 5 & 5 & 0 & 0 & 0 & 0 & 0 & 0 \\ 9 & 9 & 9 & 5 & 7 & 5 & 0 & 0 & 0 & 0 & 0 & 0 \\ 9 & 9 & 9 & 5 & 5 & 7 & 0 & 0 & 0 & 0 & 0 & 0 \\ 0 & 0 & 0 & 0 & 0 & 0 & 7 & 5 & 5 & 9 & 9 & 9 \\ 0 & 0 & 0 & 0 & 0 & 0 & 5 & 7 & 5 & 9 & 9 & 9 \\ 0 & 0 & 0 & 0 & 0 & 0 & 5 & 5 & 7 & 9 & 9 & 9 \\ 0 & 0 & 0 & 0 & 0 & 0 & 9 & 9 & 9 & 7 & 5 & 5 \\ 0 & 0 & 0 & 0 & 0 & 0 & 9 & 9 & 9 & 5 & 7 & 5 \\ 0 & 0 & 0 & 0 & 0 & 0 & 9 & 9 & 9 & 5 & 5 & 7 \end{bmatrix}$$

Fig. 6.4 The partial Gibbs derivative Δ_2^{12} on G_{12}.

$$\mathbf{D}_{12} = \begin{bmatrix} 3 & 5 & 5 & 9 & 9 & 9 & 4 & 0 & 0 & 0 & 0 & 0 \\ 5 & 3 & 5 & 9 & 9 & 9 & 0 & 4 & 0 & 0 & 0 & 0 \\ 5 & 5 & 3 & 9 & 9 & 9 & 0 & 0 & 4 & 0 & 0 & 0 \\ 9 & 9 & 9 & 3 & 5 & 5 & 0 & 0 & 0 & 4 & 0 & 0 \\ 9 & 9 & 9 & 5 & 3 & 5 & 0 & 0 & 0 & 0 & 4 & 0 \\ 9 & 9 & 9 & 5 & 5 & 3 & 0 & 0 & 0 & 0 & 0 & 4 \\ 4 & 0 & 0 & 0 & 0 & 0 & 3 & 5 & 5 & 9 & 9 & 9 \\ 0 & 4 & 0 & 0 & 0 & 0 & 5 & 3 & 5 & 9 & 9 & 9 \\ 0 & 0 & 4 & 0 & 0 & 0 & 5 & 5 & 3 & 9 & 9 & 9 \\ 0 & 0 & 0 & 4 & 0 & 0 & 9 & 9 & 9 & 3 & 5 & 5 \\ 0 & 0 & 0 & 0 & 4 & 0 & 9 & 9 & 9 & 5 & 3 & 3 \\ 0 & 0 & 0 & 0 & 0 & 4 & 9 & 9 & 9 & 5 & 5 & 3 \end{bmatrix}$$

Fig. 6.5   The Gibbs derivative $\mathbf{D}_{12}$.

of Fourier transform on groups, the steps in FFT-like algorithms for Gibbs derivatives can be performed simultaneously. That approach reduces the time complexity at the price of the space complexity. Gibbs derivative on any finite group can be calculated in two steps. However, the space complexity approximates to $O(ng + g)$.

6.7   CALCULATION OF GIBBS DERIVATIVES THROUGH DDS

Both convolution and FFT-like algorithms for Gibbs derivatives are based upon the truth-vector representation of a given function f on G. Therefore, their complexity is determined by the order g of G. In practical applications that limits the use of these algorithms to functions of a relatively small number of variables. Algorithms based on MTDDs permit calculation of Gibbs derivatives of functions of a considerable number of variables.

A procedure to calculate Gibbs derivatives is based on decomposition of Gibbs derivative into the linear combination of partial Gibbs derivatives in (6.13). It is derived as a generalization of the procedure for calculation of the Fourier transform on non-Abelian groups and as a modification of the procedure for calculation of Gibbs derivatives on finite Abelian groups through DDs [32]. The procedure consists of the following steps.

Procedure for calculation of Gibbs derivatives

1. Represent f by the MTDD.

2. Determine partial Gibbs derivatives.

3. Determine the Gibbs derivative as the linear combination of partial Gibbs derivatives.

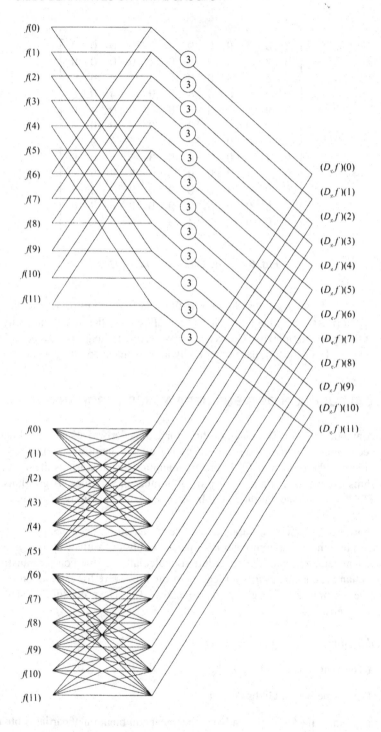

Fig. 6.6 The flow graph of the fast algorithm for calculation of the Gibbs derivative $\mathbf{D}_{12}$ on G_{12}.

The partial Gibbs derivatives are calculated through MTDD for f and represented again by MTDDs. The Gibbs derivative is determined by adding MTDDs representing the partial Gibbs derivatives.

6.7.1 Calculation of partial Gibbs derivatives

The procedure for calculation of partial Gibbs derivatives is similar to that for FFT. The difference is in the processing rules applied at the nodes and cross points in the MTDD for f. For the partial Gibbs derivative with respect to x_i, the nodes and cross points at the i-th level are processed by the rule determined by $\mathbf{D}_i$. The nodes and cross points at the other levels are processed by the rules determined by the identity matrices of the corresponding orders as determined in Definition 6.6. We assume that the x_1 is assigned to the root node, and the other variables are assigned to the other levels in the increasing order.

Procedure for calculation of partial Gibbs derivative D_i
Given a function f on the decomposable group G of the form (2.4).

1. Represent f by the MTDD.

2. Process the nodes and cross points in the MTDD in a recursive way level by level starting from the nodes at the level to which x_n is assigned up to the root node.

3. For $j = n$ to 1, process the nodes and the cross points at the j-th level by using the rule determined by $\mathbf{D}_{G_j}$ if $j = i$, $\mathbf{I}_{(g_j \times g_j)}$ if $j < i$, and $\mathbf{I}_{(K_j \times K_j)}$ if $j > i$. The output from the processing of the root node is the partial Gibbs derivative of f with respect to the variable x_i.

The procedure for calculation of the Gibbs derivatives through MTDDs is explained and illustrated by the following example.

Example 6.3 *The Gibbs derivative for functions on the group G_{24} in Example 3.5 is defined by $\mathbf{D}_f = 6\mathbf{D}_1 + 3\mathbf{D}_2 + \mathbf{D}_3$, where the partial Gibbs derivatives are given by $\mathbf{D}_1 = \mathbf{D}_{C_2} \otimes \mathbf{I}_{(2\times2)} \otimes \mathbf{I}_{(6\times6)}$, $\mathbf{D}_2 = \mathbf{I}_{(2\times2)} \otimes \mathbf{D}_{C_2} \otimes \mathbf{I}_{(6\times6)}$, $\mathbf{D}_3 = \mathbf{I}_{(2\times2)} \otimes \mathbf{I}_{(2\times2)} \otimes \mathbf{D}_{S_3}$, with*

$$\mathbf{D}_{C_2} = \begin{bmatrix} 6 & 5 \\ 5 & 6 \end{bmatrix}, \quad \mathbf{D}_{S_3} = \begin{bmatrix} 7 & 5 & 5 & 9 & 9 & 9 \\ 5 & 7 & 5 & 9 & 9 & 9 \\ 5 & 5 & 7 & 9 & 9 & 9 \\ 9 & 9 & 9 & 7 & 5 & 5 \\ 9 & 9 & 9 & 5 & 7 & 5 \\ 9 & 9 & 9 & 5 & 5 & 7 \end{bmatrix}.$$

For f in Example 3.5 calculation of the Gibbs derivative goes as follows.

To calculate the partial Gibbs derivative with respect to x_3 we perform calculations determined by definition of $\mathbf{D}_{S_3}$ at the nodes $q_{3,0}$, $q_{3,1}$, $q_{3,3}$ and the cross point $q_{3,2}$. In the nodes at the levels corresponding to x_2 and x_1, we perform the identical mapping defined by $\mathbf{I}_{(2\times2)}$.

To calculate the partial Gibbs derivative with respect to x_2, we perform the identical mapping determined by $\mathbf{I}_{(6\times6)}$ at the nodes and cross point at the level for x_3, the calculations determined by $\mathbf{D}_{C_2}$ at the nodes for x_2 and the identical mapping determined by $\mathbf{I}_{(2\times2)}$ at the root node.

Similar, to calculate partial Gibbs derivative with respect to x_1, we perform $\mathbf{D}_{C_2}$ at the root node, while at the other nodes and the cross points the identical mappings $\mathbf{I}_{(2\times2)}$ are performed.

1. *Partial Gibbs derivative with respect to x_3:*

$$q_{3,0} = \mathbf{D}_{S_3}[0,6,2,1,0,0]^T = [5,6,9,2,0,0]^T,$$

$$q_{3,1} = \mathbf{D}_{S_3}[2,1,1,0,0,0]^T = [2,0,0,3,3,3]^T$$

$$q_{3,2} = \mathbf{D}_{S_3}[1,1,1,1,1,1]^T = [0,0,0,0,0,0]^T,$$

$$q_{3,3} = \mathbf{D}_{S_3}[1,1,1,1,2,2]^T = [7,7,7,10,1,1]^T.$$

$$q_{2,0} = \mathbf{I}_{(2\times2)}\begin{bmatrix} q_{3,0} \\ q_{3,1} \end{bmatrix} = [5,6,9,2,0,0,2,0,0,3,3,3]^T,$$

$$q_{2,1} = \mathbf{I}_{(2\times2)}\begin{bmatrix} q_{3,2} \\ q_{3,3} \end{bmatrix} = [0,0,0,0,0,0,7,7,7,10,3,3]^T.$$

$$\mathbf{D}_3 = \mathbf{I}_{(2\times2)}\begin{bmatrix} q_{2,0} \\ q_{2,1} \end{bmatrix}$$

$$= [5,6,9,2,0,0,2,0,0,3,3,3,0,0,0,0,0,0,7,7,7,10,1,1]^T.$$

2. *Partial Gibbs derivative with respect to x_2:*

$$q_{3,0} = \mathbf{I}_{(6\times6)}[0,6,2,1,0,0]^T = [0,6,2,1,0,0]^T,$$

$$q_{3,1} = \mathbf{I}_{(6\times6)}[2,1,1,0,0,0]^T = [2,1,1,0,0,0]^T,$$

$$q_{3,2} = \mathbf{I}_{(6\times6)}[1,1,1,1,1,1]^T = [1,1,1,1,1,1]^T,$$

$$q_{3,3} = \mathbf{I}_{(6\times6)}[1,1,1,1,2,2]^T = [1,1,1,1,2,2]^T.$$

$$q_{2,0} = \mathbf{W}(1)\begin{bmatrix} q_{3,0} \\ q_{3,1} \end{bmatrix} = \begin{bmatrix} 6q_{3,0} + 5q_{3,1} \\ 5q_{3,0} + 6q_{3,1} \end{bmatrix}$$

$$= [10,8,6,6,0,0,1,3,5,5,0,0]^T,$$

$$q_{2,1} = \mathbf{W}(1)\begin{bmatrix} q_{3,2} \\ q_{3,3} \end{bmatrix} = \begin{bmatrix} 6q_{3,2} + 5q_{3,2} \\ 5q_{3,2} + 6q_{3,0} \end{bmatrix}$$

$$= [0,0,0,0,5,5,0,0,0,0,6,6]^T.$$

$$\mathbf{D}_2 = \mathbf{I}_{(2\times2)} \begin{bmatrix} q_{2,0} \\ q_{2,1} \end{bmatrix}$$

$$= [10, 8, 6, 6, 0, 0, 1, 3, 5, 5, 0, 0, 0, 0, 0, 0, 5, 5, 0, 0, 0, 0, 6, 6]^T.$$

3. *Partial Gibbs derivative with respect to x_1:*

   $q_{3,0}$, $q_{3,1}$, $q_{3,2}$, and $q_{3,3}$ are as in calculation of D_2.

$$q_{2,0} = \mathbf{I}_{(2\times2)} \begin{bmatrix} q_{3,0} \\ q_{3,1} \end{bmatrix} = [0, 6, 2, 1, 0, 0, 2, 1, 1, 0, 0, 0]^T$$

$$q_{2,1} = \mathbf{I}_{(2\times2)} \begin{bmatrix} q_{3,2} \\ q_{3,3} \end{bmatrix} = [1, 1, 1, 1, 1, 1, 1, 1, 1, 1, 2, 2]^T.$$

$$\mathbf{D}_1 = \mathbf{W}(1) \begin{bmatrix} q_{2,0} \\ q_{2,1} \end{bmatrix} = \begin{bmatrix} 6q_{2,0} + 5q_{2,1} \\ 5q_{2,0} + 6q_{2,1} \end{bmatrix}$$

$$= [5, 8, 6, 0, 5, 5, 6, 0, 0, 5, 10, 10, 6, 3, 5, 0, 6, 6, 5, 0, 0, 6, 1, 1]^T.$$

Therefore,

$$\mathbf{D}_{24} = 6\mathbf{D}_1 + 3\mathbf{D}_2 + \mathbf{D}_3$$

$$= [10, 1, 8, 9, 8, 8, 8, 9, 4, 4, 8, 8, 3, 7, 8, 0, 7, 7, 4, 7, 7, 2, 3, 3]^T.$$

Each step of calculation can be represented through MTDDs. For example, Figure 6.7 shows calculation of the partial Gibbs derivative with respect to x_3 for f.

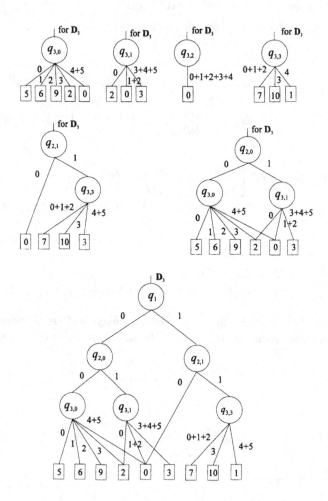

Fig. 6.7 Calculation of the partial Gibbs derivative with respect to x_3 for f in Example 3.5.

REFERENCES

1. Burrus, C.S., Eschenbacher, R.W., "An in-place, in-order prime factor FFT algorithm", *IEEE Trans. Acoust., Speech, Signal Processing*, Vol. ASSP-29, 1981, 806-816.

2. Butzer, P.L., Engels, W., Wipperfürth, U., "An extension of the dyadic calculus with fractional order derivatives: general theory," *Comp. and Math., with Appls.*, 12B, No. 5/6, 1986, 1073-1090.

3. Butzer, P.L., Stanković, R.S., Eds., *Theory and Applications of Gibbs Derivatives*, Matematički institut, Beograd, 1990.

4. Butzer, P.L., Wagner, H.J., "Walsh-Fourier series and the concept of a derivative", *Applicable Analysis*, Vol. 3, 29-46, 1973.

5. Clarke, E.M., Zhao, X., Fujita, M., Matsunaga, Y., McGeer, R., "Fast Walsh transform computation with Binary Decision Diagram", in Kebschull, U., Schubert, E., Rosentiel, W., Eds., *Proc. IFIP WG 10.5 Workshop on Applications of the Reed-Muller Expansion in Circuit Design*, 16.-17.9.1993, Hamburg, Germany, 82-85.

6. Cohn-Sfetcu, S., Gibbs, J.E., "Harmonic differential calculus and filtering in Galois fields, *Proc. 1976 IEEE Conf. on Acoustics, Speech and Signal Processing*, Philadelphia, 148-153, 1976.

7. Gibbs, J.E., "Walsh spectrometry, a form of spectral analysis well suited to binary digital computation", *NPL DES Rept.*, Teddington, Middlesex, United Kingdom, July 1967.

8. Gibbs, J.E., "Walsh functions as solutions of a logical differential equation", *National Physical Lab.*, Teddington, England, DES Report, No. 1, 1969.

9. Gibbs, J.E. Millard, M.J., "Some methods of solution of linear ordinary logical differential equations", NPL DES Rept., No. 2, Dec. 1969.

10. Gibbs, J.E., Simpson, J., "Differentiation on finite Abelian groups", *National Physical Lab.*, Teddington, England, DES Rept, No.14, 1974.

11. Gibbs, J.E., Stanković, R.S., "Matrix interpretation of Gibbs derivatives on finite groups", private communication 1989.

12. Gibbs, J.E., Stanković, R.S., "Why IWGD - 89? a look at the bibliography on Gibbs derivatives", in Butzer, P.L., Stanković, R.S., Eds., *Theory and Applications of Gibbs Derivatives*, Matematički institut, Beograd, 1990, xi-xxiv.

13. Good, I.J., "The interaction algorithm and practical Fourier analysis", *J. Roy. Statist. Soc.*, ser. B, Vol. 20, 1958, 361-372, Addendum, Vol. 22, 1960, 372-375.

14. Karpovsky, M.G., "Fast Fourier transforms on finite non-Abelian groups", *IEEE Trans. Comput.*, Vol. C-26, No. 10, 1028-1030, 1977.

15. Karpovsky, M.G., Trachtenberg, E.A., "Fourier transform over finite groups for error detection and error correction in computation channels", *Inform. Control*, Vol. 40, No. 3, 335-358, 1979.

16. Kečkić, J.D., "On some classes of linear equations", *Publ. Inst. Math. (Beograd)*, 24, (38), 1978, 89-97.

17. Le Dinh, C.T., Le, P., Goulet, R., "Sampling expansions in discrete and finite Walsh-Fourier analysis", *Proc. 1972 Symp. Applic. Walsh Functions*, Washington, DC, 1972, 265-271.

18. Moraga, C., "Introduction to linear *p*-adic invariant systems, in *Cybernetics and Systems Research*, Vol. 2, 121-124, Ed., R. Trappl, Vienna: Electronic Science Publ., 1984.

19. Nussbaumer, H.J., *Fast Fourier Transform and Convolution Algorithms*, Springer-Verlag, Berlin, Heidelberg, New York, 1981.

20. Onneweer, C.W., "Fractional differentiation on the group of integers of a p-adic or p-series field, *Anal. Math.*, Vol. 3, 119-130, 1977.

21. Schipp, F., Wade, W.R., "A fundamental theorem of dyadic calculus for the unit square", *Applicable Analysis*, Vol. 34, 1989, 203-218.

22. Stanković, R.S., "A note on differential operators on finite non-Abelian groups, *Applicable Analysis*, Vol. 21, 31-41, 1986.

23. Stanković, R.S., *Differential Operators on Groups*, Ph.D.Thesis, Faculty of Elec. Engng., Univ. Belgrade, iii+127pp, 1986 (in Serbian, summary in English 8pp).

24. Stanković, R.S., "A note on differential operators on finite Abelian groups", *Cybernetics and Systems*, 18, 221-231, 1987.

25. Stanković, R.S., "A note on spectral theory on finite non-Abelian groups", in *Theory and Applications of Spectral Techniques*, (Ed.), C. Moraga, Forschungsbericht 268, ISSN 0933-6192, Dortmund University, 1988.

26. Stanković, R.S., "Gibbs derivatives on finite non-Abelian groups", in Butzer, P.L., Stanković, R.S., (Eds.), *Theory and Applications of Gibbs Derivatives*, Matematički institut, Beograd, 1990, 269-297.

27. Stanković, R.S., "Matrix interpretation of fast Fourier transform on finite non-Abelian groups", *Res. Rept. in Appl. Math.*, YU ISSN 0353-6491, Ser. Fourier Analysis, Rept. No.3, April 1990, 1-31, ISBN 86-81611-03-8.

28. Stanković, R.S., "Matrix interpretation of the fast Fourier transforms on finite non-Abelian groups", *Proc. Int. Conf. on Signal Processing, Beijing / 90*, 22.-26.10.1990, Beijing, China, 1187-1190.

29. Stanković, R.S., "Fast algorithms for calculation of Gibbs derivatives on finite groups," *Approx. Theory and Its Applications*, 7, 2, June 1991, 1-19.

30. Stanković, R.S., "Gibbs derivatives", *Numerical Functional Analysis and Optimization*, 15, 1-2, 1994, 169-181.

31. Stanković, R.S., Stanković, M.S., "Sampling expansions for complex-valued functions on finite Abelian groups", *Automatika*, 25, 3-4, 147-150, 1984.

32. Stanković, R.S., Stanković, M., "Calculation of Gibbs derivatives through decision diagrams", *Approximation Theory and Its Applications*, China, 1998.

33. Stanković, R.S., Stojić, M.R., "A note on the discrete Haar derivative", *Colloquia Mathematica Societatis Janos Bolyai 49. Alfred Haar Memorial Conference*, Budapest, Hungary, 897-907, 1985.

34. Stanković, R.S., Stojić, M.R., "A note on the generalized discrete Haar derivative", *Automatika*, 28, 3-4, 117-122, 1987.

35. Stojić, M.R., Stanković, M.S., Stanković, R.S., *Discrete Transforms in Application*, Second Edition, Nauka, Beograd, 1993 in Serbian.

36. Thomas, L.H., "Using a computer to solve problems in physics", in *Application of Digital Computers*, Ginn, Boston, 1963.

37. Watari, Ch., "Multipliers for Walsh-Fourier series", *Tôhoky Math. J.*, 2, 16, 239-251, 1964.

38. Weiyi Su, "Gibbs derivatives and their applications," *Reports of the Institute of Mathematics, Nanjing University*, 91-7, Nanjing, P.R. China, 1991, 1-20.

39. Weiyi, Su, "Walsh analysis in the last 25 years", *Proc. The Fifth Int. Workshop on Spectral Techniques*, 15.-17.3.1994, Beijing, China, 117-127.

40. Zelin, He, "The derivatives and integrals of fractional order in Walsh-Fourier analysis, with applications to approximation theory", *J. Approx. Theory*, 39, 361-373, 1983.

7

Linear Systems and Gibbs Derivatives on Finite Non-Abelian Groups

Linearity is a property very often used in providing mathematical models of physical phenomena. In that setting, linear shift invariant systems and in particular, linear convolution systems on groups are efficiently used in mathematical modeling of real life systems. For example, in that general ground linear time-invariant systems can be regarded as systems on the real group R. Similarly, linear discrete-time-invariant systems are an example of systems defined on the additive group of integers Z. The use of some other groups different from R and Z offers some advantages in particular applications, see for example [13], [14].

Differential operators are used in linear systems theory to describe the change of state of a system. The systems on R described by linear differential equations with constant coefficients in terms of the Newton-Leibniz derivative are probably the most familiar example. However, group theoretic models of systems and Gibbs derivatives on groups, in particular on the dyadic and groups Cp^n, have attained some considerable attention [18], [15], [8].

In what follows we will first give a short account of background to linear systems on groups and then discuss systems on finite non-Abelian groups described by discrete differential equations using Gibbs derivatives.

7.1 LINEAR SHIFT-INVARIANT SYSTEMS ON GROUPS

In this section, we briefly discuss linear convolution systems with whose input and output signals are deterministic signals on groups, and point out the relationship between Gibbs differentiators and linear convolution systems.

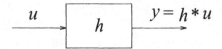

Fig. 7.1   Linear shift-invariant system.

Definition 7.1 *A linear invariant system S over a group G is defined as a quadruple $S = (U, Y, h, *)$ where the operation $*$ is defined for any $u \in U$, $y \in Y$ as follows:*

$$y(t) = (h * u)(t) = \int_{x \in G} h(x^{-1} \circ t)u(x), \qquad (7.1)$$

i.e., $*$ is the operation of group convolution of two functions h, u; x^{-1} is the inverse of x in G and $\circ$ denotes the group operation.

Worded differently, the system S consists of the set U of input signals and the set Y of output signals defined respectively, as the mappings $u : G \to X$ and $y : G \to Y$, and the impulse function h defined as the mapping $h : U \to Y$. If (7.1) is true for a given system S, and given $u \in U, y \in Y$, then that system computes the input/output pair (u, y).

Figure 7.1 shows a general model of a linear shift-invariant system on groups.

Note that from the system theory point of view, S is a linear input/output system whose input and output are defined over an arbitrary group G. By using different groups, various systems studied by several authors can be obtained. For example, if G is the dyadic group, the dyadic systems ware introduced by Pichler and further studied in a series of papers by this and by several other authors; see [12] for a bibliography up to 1989. For more recent result, we refer to [7], [8], [21].

The systems where input and output signals are modeled by functions mapping infinite cyclic group of integers into Galois fields of order 2^q, $q \in N, GF(2^q)$ were considered by Tsypkin and Faradzhev [32]. A generalization of the concept was given in [23] where it was shown that both cyclic and dyadic convolution systems on finite groups can be regarded as special classes of permutation-invariant systems. In a more general setting, systems over locally compact Abelian groups were considered by Falb and Friedman [9]. Some aspects of the theory were extended also to non-Abelian groups by Karpovsky and Trachtenberg [13], [14].

Recall that systems over finite groups can be regarded as a special class of digital filters [28], [29] [30] or a special class of discrete-time systems with variable structure [16] over a finite interval $[0, g - 1]$, see also [31].

It may be said that in last few years the theory of linear invariant systems on groups has been well established by several authors, although a lot of further work is still needed. Regarding the applications of these systems note that they can be used as models of both information channels (for example to represent an encoder, a digital filter, or a Wiener filter if K is the field of complex numbers and u is a stochastic signal), and computation channels if K is a finite field. For example, different criteria for the approximation of linear time-invariant systems by linear convolution systems

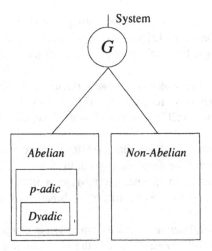

Fig. 7.2 Classification of linear shift-invariant systems on groups.

on groups were discussed in [31]. Figure 7.2 shows a classification of linear shift-invariant systems with respect to the domain groups for input and output signals.

7.2 LINEAR SHIFT-INVARIANT SYSTEMS ON FINITE NON-ABELIAN GROUPS

In the case of systems on finite non-Abelian groups the Definition 7.1 can be stated as follows.

Definition 7.2 *A scalar linear system A over a finite not necessarily Abelian group G is defined as a quadruple* $(P(G), P(G), h, *)$ *where the input-output relation* $*$ *is the convolution product on G,*

$$y = h * f, \quad f, h, y \in P(G),$$

i.e.,

$$y(\tau) = \sum_{x \in G} h(x) f(\tau x^{-1}), \forall \tau \in G. \tag{7.2}$$

So, an ordered pair $(f, y) \in P(G) \times P(G)$ is exactly then an input-output pair of A if f and y fulfill equation (7.2). The function $h \in P(G)$ is the impulse response of A.

It is easy to show that the system A is invariant against the translation of input function. By that we mean that if y is the output to f, then $T^\tau y$ is the output to $T^\tau f$, for all $\tau \in G$. Therefore, we denote the system A as a linear translation invariant (LTI) system.

It is apparent that when G is the dyadic group, the Definition 7.2 reduces to the dyadic systems introduced in [17] and further studied in [18] and a series of papers of that and other authors. If G is the group Z_{p^n} we obtain the systems studied in [4], [6], and [15].

The dyadic and p-adic systems are closely related with Gibbs differentiators on the corresponding groups see, for example [15], [18]. A corresponding relationship can be established between LTI systems and Gibbs derivatives on finite non-Abelian groups [24].

First of all, note that (6.5) shows that the Gibbs differentiator D^k of order k is a LTI system having an impulse response h given by $h = \delta^{(k)}$, see [24].

The Gibbs discrete differential equation (6.6) can be interpreted as an input-output relation of a system A belonging to a linear combination of Gibbs derivatives on a finite non-Abelian group.

From (6.7), the general output function of this system is represented as the sum of the zero-input response of the system y_{zi} and the zero-state response y_{zs} has the form identical to (7.2). Therefore, we infer that the scalar linear system A associated with (7.2) is a LTI system for which (6.7) represents an input-output-state relation and h is the impulse response of A to the unit impulse $\delta(x)$. Since h is the inverse Fourier transform of $H(w)$, the transfer function of A is given by (6.9).

7.3 GIBBS DERIVATIVES AND LINEAR SYSTEMS

The relationship between linear convolution systems on locally compact Abelian and finite non-Abelian groups discussed above can be considered and summarized in a general setting as follows.

In a general ground the Gibbs differentiator of order k of a function $f \in K(G)$, which we denote by $D^k f$, is considered as the linear operator in $K(G)$ satisfying the relationship [27]

$$(F(D^k f))(w) = \varphi(w, k)(F(f))(w),\tag{7.3}$$

where F denotes the Fourier transform operator in $K(G)$.

In most cases $\varphi(w, k) = w^k$, but in some cases a scaling factor should be applied, while in a few particular cases the function φ differs and is related to the order of group G. For example, in the case of the extended Butzer-Wagner dyadic derivative [1] $\varphi(w, k) = (w^*(w))^k$, where

$$w^*(w) = \sum_{i=0}^{\infty} (-1)^i w_i 2^i,$$

w_i being the coefficients in the dyadic expansion of $w \in P$. In the case of Gibbs derivatives on Vilenkin groups [35], [36], [37], the function $\varphi(w, k)$ is a function from the so-called symbol class $S_{\rho,\sigma}^m$ [35] defined as $\varphi(w, k) = \langle k \rangle^m, m \geq 0$, where $\langle x \rangle = \max\{1, \|x\|\}$.

It should be pointed out in attempting to determine a relationship between Gibbs derivatives and linear convolution systems that

1. Thanks to the relation (7.3) and the convolution theorem in the Fourier analysis on groups, the Gibbs differentiator of order k can be considered as a convolution operator and, therefore, can be identified with a linear convolution system whose impulse response function h is given in the transform domain by $(F(h))(w) = \varphi(w, k)$. For example, in the case of the Gibbs derivative on finite not necessarily Abelian groups, as well as in the case of dyadic and groups Cp^n, $\varphi(w, k) = w^k$ by definition and, therefore, h is the k-th Gibbs derivative of the δ-function defined as $\delta(x) = 1$ for x equals the unit element of G, and $\delta(x) = 0$ otherwise.

2. A considerable class of linear systems on groups can be described by Gibbs differential equations in a way resembling the use of classical differential equations with constant coefficients in the linear system theory on the real group R. In the other words, a linear Gibbs differential on a group is defined as a polynomial in the Gibbs differentiator with real coefficients. The linear Gibbs differential operators form a subset of the group convolution operators realized by the corresponding subset of the group convolution systems.

As we noted above such linear systems over dyadic groups were discussed in [11], [22], and for the infinite dyadic groups in [18]. Recall that an extension of the theory to p-adic groups, the finite Vilenkin groups, was given in [6], [15]. A generalization to finite non-Abelian groups was given in [24], see also [26] and for p-adic systems with stochastic signals in [8].

Note that the use of systems modeled by Gibbs differential equations in the processing of two-dimensional signals was suggested in [19], [20].

7.3.1 Discussion

As is known, the dyadic derivative is especially adapted to functions having many jumps and possessing just a few and short intervals of constancy. Even functions having a denumerable set of discontinuities like the well-known Dirichlet function can be dyadically differentiable on $[0, 1]$. In the case of finite groups, the Gibbs derivatives also provide a mean to differentiate functions on those groups. In one word, through the family of Gibbs differentiators, the advantage of the use of differential calculus extends to the theory of systems whose input/output signals are piecewise constant, or discrete functions.

In order to point out some possible advantages of linear systems on groups modeled by the Gibbs differential equations, recall that the use of Fourier analysis in linear systems theory is based upon the convolution theorem and the relationship between the Newton-Leibniz derivative. Thanks to the first property, the Fourier transform maps the convolution into ordinary multiplication, while the second permits the translation of differential equations into the algebraic ones. As in many other areas, the application of Fourier analysis in linear systems theory is further supported by the existence of the fast Fourier transform, FFT, and related algorithms for efficient calculation of Fourier coefficients and some other parameters useful in practical applications.

The Gibbs derivatives possesses the most of the useful properties of Newton-Leibniz derivative, except the product rule and, therefore, their role in the theory of linear systems on groups can be compared to that of Newton-Leibniz derivative in classical linear systems theory on R. At the same time, the Gibbs derivatives are efficiently characterized by the Fourier coefficients on groups.

The matrices representing Gibbs derivatives are Kronecker product representable in the case of finite decomposable groups, (see Section 4.7), and, therefore, the fast algorithms for the calculation of the values of Gibbs derivatives on these groups can be defined, (see Section 4.8).

It may be said that the Gibbs differentiation shares some of very useful properties of both Fourier analysis and differential calculus.

Thanks to these properties the Gibbs derivatives could be very promising for the use in the theory of linear systems on groups. Some recent results and extensions of the theory are given in [7], [21].

GIBBS DERIVATIVES AND LINEAR SYSTEMS

REFERENCES

1. Butzer, P.L., Engels, W., Wipperfürth, U., "An extension of the dyadic calculus with fractional order derivatives: general theory," *Comp. and Math., with Appls.*, 12B, No. 5/6, 1986, 1073-1090.

2. Butzer, P.L., Wagner, H.J., "Walsh-Fourier series and the concept of a derivative", *Applicable Analysis*, Vol. 1, No. 3, 29-46, 1973.

3. Butzer, P.L., Stanković, R.S., Eds., *Theory and Applications of Gibbs Derivatives*, Matematički institut, Beograd, 1990.

4. Cohn-Sfetcy, S., "On the theory of linear dyadic invariant systems", *Proc. Symp. Theory and Applic. Walsh and Other Non-sinus. Functions*, Hatfield, England, 1975.

5. Cohn-Sfetcy, S., *Topics on generalized convolution and Fourier transform: Theory and applications in digital signal processing and system theory*, Ph.D. Thesis, McMaster University, Hamilton, Ontario, Canada, 1976.

6. Cohn-Sfetcu, S., Gibbs, J.E., "Harmonic differential calculus and filtering in Galois fields," *Proc. IEEE Int. Conf. Acoust. Speech and Signal Processing*, Philadelphia, 1976 April 12-14, 148-153.

7. Endow, Y., "Walsh harmonizable processes in linear system theory", *Cybernetics and Systems*, 1996, 489-512.

8. Endow, Y., Stanković, R.S., "Gibbs derivatives in linear system theory", in: R. Trappl, Ed., *Proc. Twelfth European Meeting on Cybernetics and Systems Research*, 5.-8.4.1994, Vienna, Austria.

9. Falb, P.L., Friedman, M.I., "A generalized transform theory for causal operators", *SIAM J. Control*, 1970, 452-471.

10. Gibbs, J.E., Simpson, J., "Differentiation on finite Abelian groups", *NPL DES Rept.*, No. 16, July 1974.

11. Gibbs, J.E., Marshall, J.E., Pichler, F.R., "Electrical measurements in the light of system theory", *IEE Conference on the Use of Computers in Measurement*, University of York, September 24-27, 1973, ii+5pp.

12. Gibbs, J.E., Stanković, R.S., "Why IWGD- 89? A look at the bibliography of Gibbs derivatives," in: *Theory and Applications of Gibbs Derivatives*, P.L. Butzer, R.S. Stanković, Eds., Matematički institut, Beograd, 1990, $xi - xxiv$.

13. Karpovsky, M.G., Trachtenberg, E.A., "Some optimization problems for convolution systems over finite groups", *Inf. and Control*, 34, 1977, 227-247.

14. Karpovsky, M.G., Trachtenberg, E.A., "Statistical and computational performances of a class of generalized Wiener filters", *IEEE Trans. Inform. Theory*, Vol. IT-32, No. 2, 1986, 303-307.

15. Moraga, C., "Introduction to linear p-adic systems," R. Trappl, Ed., *Cybernetics and Systems Research, 2,* North-Holland, Amsterdam, 1984.

16. Nailor, A.W., "A transform technique for multivariable, timevarying, discrete-time linear systems", *Automatica*, 2, 1965, 211-234.

17. Pearl, J., "Optimal dyadic models of time-invariant systems", *IEEE Trans. Computers*, Vol. C-24, 1975.

18. Pichler, F., "Walsh functions and linear system theory", Proc. Applic. Walsh Functions, Washington, DC, 1970, 175-182.

19. Pichler, F., "Fast linear methods for image filtering", in: D.G. Lainiotis, N.S. Tzanues, Eds., *Applications of Information and Control Systems*, D. Reidel Publishing Company, Dordrecht, 1980, 3-11.

20. Pichler, F., "Experiments with 1-D and 2-D signals using Gibbs derivatives", in: *Theory and Applications of Gibbs Derivatives*, P.L. Butzer, R.S. Stanković, Eds., Matematički institut, Beograd, 1990, 181-196.

21. Pichler, F., "Realizations of Priggogine's Λ-transform by dyadic convolution", in Trappl, R., Horn, W., (eds.), Austrian Society for Cybernetic Studies, ISBN 385206127X, 1992.

22. Pichler, F.R., Marshall, J.E., Gibbs, J.E., "A system-theory approach to electrical measurements", Colloquium on the Theory and Applications of Walsh Functions, The Hatfield Polytechnic, Hatfield, Hertfordshire, June 28-29, 1973.

23. Siddiqi, M.U., Sinha, U.P., "Permutation-invariant systems and their application in the filtering of finite discrete data", *Proc. IEEE Int. Conf. on Acoustics, Speech, and Signal Processing, ICASSP'77*, May 1977, Vol. 2, 352-355.

24. Stanković, R.S., "Linear harmonic translation invariant systems on finite non-Abelian groups", R. Trappl, Ed., *Cybernetics and Systems Research* North-Holland, Amsterdam, 1986.

25. Stanković, R.S., "A note on differential operators on finite non-Abelian groups," *Applicable Anal.*, 21, 1986, 31-41.

26. Stanković, R.S., "A note on spectral theory on finite non-Abelian groups," *3rd Int. Workshop on Spectral Techniques*, Univ. Dortmund, 1988, Oct.4-6, 1988.

27. Stanković, R.S., "Gibbs derivatives", *Numerical Functional Analysis and Optimization*, 15, 1-2, 1994, 169-181.

28. Trachtenberg, E.A., "Systems over finite groups as suboptimal filters: a comparative study", P.A. Fuhrmann, Ed., *Proc. 5th Int. Symp. Math. Theory of Systems and Networks*, Springer-Verlag, Beer-Sheva, Israel, 1983, 856-863.

29. Trachtneberg, E.A., "Fault tolerant computing and reliable communication: a unified approach", *Information and Computation*, Vol. 79, No. 3, 1988, 257-279.

30. Trachtenberg, E.A., Karpovsky, M.G., "Optimal varying dyadic structure models of time invariant systems", *Proc. 1988 IEEE Int. Symp. Circuits and Systems*, Espoo, Finland, June 1988.

31. Trachtenberg, E.A., "SVD of Frobenius matrices for approximate and multiobjective signal processing tasks", in: E.F. Deprettere, Ed., *SVD and Signal Processing*, Elsevier North-Holland, Amsterdam/New York, 1988, 331-345.

32. Tsypkin, Ya.Z., Faradzhev, R.G., "Laplace-Galois transformation in the theory of sequential machines", *Dokl. Akad. Nauk USSR*, Vol. 166, No. 3, 1966, 507-573 (in Russian).

33. Wade, W.R., "Decay of Walsh series and dyadic differentiation", *Trans. Amer. Math. Soc.*, Vol. 277, No. 1, 413-420, 1983.

34. Watari, Ch., "Multipliers for Walsh-Fourier series", *Tôhoky Math. J.*, 2, 16, 239-251, 1964.

35. Weiyi, Su., "Pseudo-differential operators in Bessov spaces over local fields", *J. Approx. Theory and Appls.*, Vol. 4, No. 2, 1988, 119-129.

36. Weiyi Su, "Gibbs derivatives and their applications," *Reports of the Institute of Mathematics, Nanjing University*, 91-7, Nanjing, P.R. China, 1991, 1-20.

37. Weiyi Su, "Pseudo-differential operators and derivatives on locally compact Vilenkin groups," *Science in China* (Series A), Vol. 35, No. 7, 1992, 826-836.

8

Hilbert Transform on Finite Groups

In theory of real variable functions the Hilbert transform is defined in the following way.

Definition 8.1 *The Hilbert transform $f^{\sim}$ of a function $f \in L^p, 1 \neq p \neq \infty$ is defined, see for example [2], [9], by*

$$f^{\sim}(x) = v.p. \left(\frac{1}{\pi} \int_{-\infty}^{\infty} \frac{f(x-u)}{u} du \right), \tag{8.1}$$

where the notation v.p. means that the integral is understood in the sense of Cauchy principal value.

The functions belonging to L^2 are the most widely exploited in practice, since they represent the finite energy signals. The following holds for these functions, see for example [2], [9].

Denote by $F(w)$ the Fourier transform of a function $f \in L^2$. Then, the Fourier transform of its Hilbert transform, $F^{\sim}(w)$, is given by

$$F^{\sim}(w) = -\text{sign}(w)F(w) \quad a.e., \tag{8.2}$$

where

$$\text{sign}(w) = \begin{cases} -1, & w < 0, \\ 0, & w = 0, \\ 1, & w > 0. \end{cases} \tag{8.3}$$

The relation (8.3) can be regarded as an alternative definition of the Hilbert transform. The formula holds for $p = 1$ if $f^{\sim} \in L^1$, in which case it holds everywhere.

Note that there are functions from l^1 whose Hilbert transforms defined by (8.2) do not belong to l^1. An example is $f(x) = \frac{1}{1+x}$ as is noted in [2].

Recall that

$$v.p.\left(\frac{1}{\sqrt{2\pi}}\int_{-\infty}^{\infty}\exp\left(\frac{-iwx}{x}dx\right)\right) = \lim_{\epsilon\to 0}\frac{-2i}{\sqrt{2\pi}}\int_{\epsilon}^{\infty}\frac{\sin(wx)}{x}dx$$

$$= \{-i\text{sign}(w)\}\sqrt{\frac{2}{\pi}}\int_{0}^{\infty}\frac{\sin y}{y}dy = \{-i\text{sign}(w)\}\sqrt{\frac{\pi}{2}}.$$

From there it can be written at least formally:

$$F^{\sim}(w) = \mathbf{F}\left(\sqrt{\frac{2}{\pi}}\left(f*\frac{1}{\pi}\right)\right) = \{-\text{sign}(w)\}\mathbf{F}(w), \tag{8.4}$$

where $\mathbf{F}$ is the Fourier transform operator, $*$ denotes the convolution product on R, and the convolution integral is understood in the sense of Cauchy principal value.

In the case of periodic functions with the period equal to 2π the convolution kernel used to define the corresponding Hilbert transform is $\{\cot\frac{x}{2}\}$.

The approach of defining the Hilbert transform in transform domain as the multiplication by a sign function, that is by employing the relation (8.3) was used as the starting point for the introduction of a discrete Hilbert transform, i.e., the Hilbert transform for functions on finite Abelian groups. It is important to note that definitions appearing in [3] and [1] [4], [5], [6] are based upon the differently defined sign functions and they coincide only in the case of cyclic groups.

Recalling that the real line R exhibits the structure of a locally compact Abelian group, it can be concluded that the Hilbert transform for real-variable functions and the discrete Hilbert transform can be considered uniformly as the Hilbert transform on Abelian groups. However, the above discussed group-theoretic approach of introducing the Hilbert transform on Abelian groups through the product in the transform domain by a suitably defined sign function, can hardly be used further to extend the concept to finite non-Abelian groups. That fact becomes obvious if we recall that unlike Abelian groups, the domain Γ of the Fourier transform S_f of a function f on a finite not necessarily Abelian group G may not have any algebraic structure suitable to define a multiplication in it which in turn can be mapped into a convolution in the group. Therefore, we have suggested in [8] just the opposite way, we have defined a Hilbert transform on a finite non-Abelian group as the pointwise multiplication of a given function by a suitably defined sign function in the group. As was shown in [8], an analysis of the properties of the thus defined transform justifies to consider it as a proper counterpart of the Hilbert transform on R or on finite Abelian groups. Therefore, we are encouraged to suggest this "opposite way to be actually a "proper" way to define a Hilbert-like transform for functions on both Abelian and non-Abelian groups permitting the considerations of these two cases in a uniform way. Recall that the same approach was already used for Hilbert transform on R in some particular engineering applications as for example signal filtering.

In that way two aims are reached. First, the main properties of the "classically" defined Hilbert transform on Abelian groups are preserved by the "new" transform, and the concept is extended to finite non-Abelian groups. Further, as it will be shown below, the same approach can be used to introduce a Hilbert-like transform for functions mapping a given finite non-Abelian group into a finite field admitting the existence of a Fourier transform.

8.1   SOME RESULTS OF FOURIER ANALYSIS ON FINITE NON-ABELIAN GROUPS

For the sake of completeness of presentation in this section we disclose several further results of Fourier analysis on finite non-Abelian groups which are somewhat restricted counterparts of the corresponding results on finite Abelian groups. Recall again that the Fourier transform S_F is defined on Γ and, thus, cannot be regarded as a function on a group and, therefore, some of the properties valid on Abelian groups are non-existent when the group is no longer commutative.

Recalling the bijection V from non-Abelian group G of order g onto the subset $M = \{0, \ldots, g - 1\}$ of integers adopted in this monograph, note that the natural ordering "$<$" in M induces an ordering upon G via the inverse mapping V^{-1}. We keep the symbol "$<$" for the new ordering in G and define the following partition of G:

$$POS_V = \{x \in G | x < x'\}, \quad SYM_V = \{x \in G | x = x'\}, \quad NEG_V\{x \in G | x' < x\},$$

where x' is the inverse of x in G, i.e., $x \circ x' = e$ and $(x')' = x$. The symbols $\circ$ and e represent the group operation and the identity of G. Notice that both the set POS and NEG depend on the bijection V.

Among the functions from $P(G)$ we will not the following special classes.

Definition 8.2 *Let* $f \in P(G)$. *Then, with respect to the ordering of G introduced by V, we say:*

1. *f is even iff $f(x) = f(x'), \forall x \in G$,*

2. *f is odd iff $f(x) = -f(x'), \forall x \in G$,*

3. *f is actual iff $f(x) = 0, \forall x \in G \setminus POS_V$,*

4. *f is coactual iff $f(x) = 0, \forall x \in POS_V$,*

5. *f is virtual iff $f(x) = 0, \forall x \in G \setminus NEG_V$,*

6. *f is covirtual iff $f(x) = 0, \forall x \in NEG_V$,*

7. *f is axial iff $f(x) = 0, \forall x \in G \setminus SYM_V$,*

8. *f is coaxial iff $f(x) = 0, \forall x \in SYM_V$.*

Notice that if f is odd, then $f(x) = 0, \forall x \in SYM_V$, and if f is axial, then it is also even.

Definition 8.3 *For all $x \in G$, the function $sign_V$ is defined as follows*

$$sign_V(x) = \begin{cases} 1, & x \in POS_V, \\ -1, & x \in NEG_V, \\ 0, & x \in SYM_V. \end{cases}$$

It becomes apparent that $sign_V$ is a coaxial function.

Property 8.1 *Let $f \in P(G)$. Then,*

1. *$f_e(x) = \frac{1}{2}(f(x) + f(x'))$ defines an even function,*

2. *$f_o(x) = \frac{1}{2}(f(x) - f(x'))$ defines an odd function,*

3. *$f(x) = f_e(x) + f_o(x)$.*

The sign minus in Property 8.1 as well as in Definition 8.2 is understood in the sense of subtraction in the field P. For the correctness of the notation, the derivation of the following properties will be restricted to the complex functions on G. The real functions will be considered as a subclass of these functions in which case we will use the notation $R(G)$. Note that the corresponding properties can be derived for function in finite fields, but the difference which should be appreciated is that in that case the notion of imaginary unity and, consequently, the complex conjugate does not exist in the "classical" sense. Therefore, the concepts of Hermitean and skew-Hermitean matrix should be appropriately reformulated as it will be given in the corresponding section below.

Proof. The statement is obvious from corresponding definitions.

Property 8.2 *Let $f \in R(G)$. Then f is odd iff its Fourier transform $\mathbf{S}_f$ is skew-Hermitean, i.e., iff for each $0 \neq w \neq K - 1$, $\mathbf{S}_f(w) = -\overline{\mathbf{S}}_f^T(w) = -\mathbf{S}_f^*(w)$, where $\mathbf{S}_f^T$ denotes the transpose, $\overline{\mathbf{S}}_f$ the complex-conjugate, and $\mathbf{S}_f^*$ the complex-conjugate transpose of $\mathbf{S}_f$.*

Proof. Assume first that f is odd, i.e., $f(x) = -f(x'), \forall x \in G$. Then, for $w = 0, \ldots, K - 1$, we have

$$\mathbf{S}_f(w) = r_w g^{-1} \sum_{x \in G} f(x) \overline{\mathbf{R}_w^T} = -r_w g^{-1} \sum_{x \in G} f(x') \overline{\mathbf{R}_w^T(x)}.$$

Since f is real we obtain

$$\mathbf{S}_f(w) = \overline{\left(-r_w g^{-1} \sum_{x \in G} f(x') \mathbf{R}_w \right)^T}(x) = -\mathbf{S}_f^*(w).$$

Conversely, assume that $\mathbf{S}_f$ is skew-Hermitean, i.e., $\mathbf{S}_w = -\mathbf{S}_f^*(w)$. Now, since f is real-valued, for each $x \in G$ we have

$$f(x) = \sum_{w=0}^{K-1} Tr(\mathbf{S}_f(w)\mathbf{R}_w(x)) = \overline{\sum_{w=0}^{K-1} Tr(\mathbf{S}_f(w)\mathbf{R}_w(x))}$$

$$= \sum_{w=0}^{K-1} Tr(\overline{\mathbf{S}_f(w)\mathbf{R}_w(x)}) = \sum_{w=0}^{K-1} \overline{Tr(\overline{\mathbf{S}_f(w)\mathbf{R}_w(x)})}^T$$

$$= \sum_{w=0}^{K-1} Tr(\overline{\mathbf{S}_f^T(w)\mathbf{R}_w^T(x)}) = -\sum_{w=0}^{K-1} Tr(\mathbf{S}_f(w)\overline{\mathbf{R}_w^T(x)})$$

$$= -\sum_{w=0}^{K-1} Tr(\mathbf{S}_f(w)\mathbf{R}_w(x')) = -f(x').$$

Property 8.3 *Let* $f \in R(G)$. *Then,* f *is even iff its Fourier transform* $\mathbf{S}_f$ *is Hermitean, i.e., iff for each* $0 \neq w \neq K - 1$, $\mathbf{S}_f(w) = \mathbf{S}_f^*(w)$.

Proof. First assume that f is even, i.e., $f(x) = f(x'), \forall x \in G$. Then for all $w = 0, \ldots, K - 1$

$$\mathbf{S}_f(w) = r_w g^{-1} \sum_{x \in G} f(x)\overline{\mathbf{R}_w^T(x)} = r_w g^{-1} \sum_{x \in G} f(x')\overline{\mathbf{R}_w^T(x)}$$

$$= r_w g^{-1} \sum_{x \in G} \left(\overline{f(x')\mathbf{R}_w(x)}\right) = \left(\overline{r_w g^{-1} \sum_{x \in G} f(x)\mathbf{R}_w^T(x)}\right)^T$$

$$= \overline{(\mathbf{S}_f(w))}^T.$$

Thus, $\mathbf{S}_f$ is Hermitean.

Conversely, assume that $\mathbf{S}_f$ is Hermitean, i.e., $\mathbf{S}_f(w) = \overline{\mathbf{S}_f^T(w)}$. Then, since $f \in R(G)$, we have

$$f(x) = \overline{f(x)} = \overline{\sum_{w=0}^{K-1} Tr(\mathbf{S}_f(w)\mathbf{R}_w(x))} = \sum_{w=0}^{K-1} Tr((\overline{\mathbf{S}_f(w)\mathbf{R}_w(x)}))^T$$

$$= \sum_{w=0}^{K-1} Tr(\mathbf{S}_f(w)\overline{\mathbf{R}_w^T(x)}) = \sum_{w=0}^{K-1} Tr(\mathbf{S}_f(w)\mathbf{R}_w(x'))$$

$$= f(x').$$

Property 8.4 *Let* $f \in R(G)$ *and* $f(x) = f_e(x) + f_o(x)$. *Then,*

$$\mathbf{S}_{f_e}(w) = \frac{1}{2}(\mathbf{S}_f(w) + \mathbf{S}_f^*(w)),$$

$$\mathbf{S}_{f_o}(w) = \frac{1}{2}(\mathbf{S}_f(w) - \mathbf{S}_f^*(w)),$$

for each $0 \neq w \neq K - 1$.

Proof. Because of the linearity of the Fourier transform

$$\mathbf{S}_f(w) = \mathbf{S}_{f_e}(w) + \mathbf{S}_{f_o}(w),$$
$$\mathbf{S}_f^*(w) = \mathbf{S}_{f_e}^*(w) + \mathbf{S}_{f_o}^*(w),$$

for each $0 \neq w \neq K - 1$ and with Properties 8.2 and 8.3 we obtain

$$\mathbf{S}_f^*(w) = \mathbf{S}_{f_e}(w) - \mathbf{S}_{f_o}(w),$$

from where the assertion follows directly.

From there we have that $\mathbf{S}_{f_e}$ is the Hermitean part of $\mathbf{S}_f$ and similarly, $\mathbf{S}_{f_o}$ is the skew-Hermitean part of $\mathbf{S}_f$, which we denote by $H(\mathbf{S}_f)$ and $sH(\mathbf{S}_f)$, respectively. This is a direct consequence of the linearity of Fourier transform and Properties 2 and 3.

Property 8.5 *The Fourier transform of the sign function is given by*

$$\mathbf{S}_{sign}(w) = r_w g^{-1} \sum_{x \in POS_V} (\mathbf{R}_w^*(x) - \mathbf{R}_w(x)).$$

Proof.

$$
\begin{aligned}
\mathbf{S}_{sign}(w) &= r_w g^{-1} \sum_{x \in G} sign(x) \mathbf{R}_w^*(x) \\
&= r_w g^{-1} \sum_{x \in POS_V} \mathbf{R}_w^*(x) - r_w g^{-1} \sum_{x \in NEG_V} \mathbf{R}_w^* \\
&= r_w g^{-1} \sum_{x \in POS_V} (\mathbf{R}_w^*(x) - \mathbf{R}_w^*(x')) \\
&= r_w g^{-1} \sum_{x \in POS_V} (\mathbf{R}_w^*(x) - \mathbf{R}_w(x)).
\end{aligned}
$$

Corollary 8.1   *1. Let $f \in R(G)$ be covirtual. A covirtual function has a trivial decomposition in an actual function $g_a = g_e + g_o$ and an axial function g_{sym}. Then we have*

$$sH(\mathbf{S}_f) = \mathbf{S}_{g_o}.$$

2. Let $f \in R(G)$ be coactual. A coactual function has a trivial decomposition in a virtual function $g_v = k_e + k_o$ and an axial function g_{sym}. Then we have

$$sH(\mathbf{S}_f) = \mathbf{S}_{k_o}.$$

Proof. Let $f \in R(G)$ be covirtual. Define

$$g_a(x) = \begin{cases} f(x), & \forall x \in POS_V \\ 0, & \text{otherwise.} \end{cases}$$

Obviously, g_a is actual and g_{sym} is even. Moreover, $f = g_a + g_{sym}$, i.e., $f = g_o + g_e + g_{sym}$. Since the sum of two even functions is even, then g_o represents the odd part of f and the assertion follows from property 2.

Similarly if f is a coactual function.

8.2 HILBERT TRANSFORM ON FINITE NON-ABELIAN GROUPS

As we noted above, the definition of the Hilbert transform based upon the relation (8.3) cannot be extended to functions defined on a non-Abelian group because the domain Γ of the Fourier transform $\mathbf{S}_f$ may not exhibit a suitable algebraic structure. For that reason we will use a reverse approach which leads to a definition holding uniformly for Abelian and non-Abelian groups.

Definition 8.4 *The Hilbert transform $f^{\sim}$ of a function $f \in P(G)$, where G is not necessarily an Abelian group, is defined under a given ordering bijection V as the linear operator $\sim: P(G) \rightarrow P(G)$ given by*

$$f^{\sim}(x) = -i\text{sign}_V(x)f(x), \quad \forall x \in G.$$

The main properties of the thus defined Hilbert transform are given in the following theorem which justifies to consider it as a proper counterpart of the Hilbert transform on R and on finite Abelian groups.

Theorem 8.1 *The main properties of the Hilbert transform are*

1. *For each $f, h \in C(G)$, and $a, b \in C$*
   $(af + bh))^{\sim}(x) = af^{\sim}(x) + bh^{\sim}(x).$

2. *If $f \in R(G)$, then $f^{\sim}$ is purely imaginary. Moreover, if f is even, then $f^{\sim}$ is odd, and if f is odd, $f^{\sim}$ is even.*

3. *For each $f \in C(G)$, $\overline{(f(x))^{\sim}} = -\overline{f}^{\sim}(x)$, where $\overline{f(x)}$ denotes the complex conjugate of $f(x)$.*

4. *For each $f, h \in C(G), f^{\sim}(x)h(x) = f(x)h^{\sim}(x)$*

5. *The inverse Hilbert transform of a coaxial function $f \in C(G)$ is given by $f^{\sim}(f^{\sim}(x)) = -f(x)$. Notice that the actual and virtual functions are also coaxial functions, and it follows $f(x)h(x) = -f^{\sim}(x)h^{\sim}(x)$.*

6. *Let $f_a \in R(G)$ be actual. Then the Hermitean and skew-Hermitean parts of its Fourier spectrum $\mathbf{S}_f$ are related by the Hilbert transform as shown below: For each $w \in \{0, \ldots, K-1\}$*

$$H(\mathbf{S}_{f_a}(w)) = (((isH(\mathbf{S}_{f_a})))^{\perp})^{\sim})^{\top}, \tag{8.5}$$

$$sH(\mathbf{S}_{f_a}(w)) = i(((H(\mathbf{S}_{f_a})))^{\perp})^{\sim})^{\top}. \tag{8.6}$$

Let $f \in R(G)$ be virtual. Then,

$$H(\mathbf{S}_{f_v}(w)) = -(((isH(\mathbf{S}_{f_v})))^{\perp})^{\sim})^{\top}, \tag{8.7}$$

$$sH(\mathbf{S}_{f_v}(w)) = -i(((H(\mathbf{S}_{f_v})))^{\perp})^{\sim})^{\top}, \tag{8.8}$$

where $\top$ and $\perp$ denote the direct and the inverse Fourier transform, respectively.

7. *Let $f \in R(G)$ be covirtual. Then equation (8.7) is also valid.*
   Let $f \in R(G)$ be coactual. Then equation (8.9) is also valid.

8. *Let $f \in R(G)$. Moreover, let $Ev(f)$ and $Od(f)$ denote the even and odd parts of f, respectively. These parts are related by the Hilbert transform as follows: If f is actual, then*

$$Ev(f(x)) = i(Od(f(x)))^\sim, \tag{8.9}$$

$$Od(f(x)) = i(Ev(f(x)))^\sim, \tag{8.10}$$

and if f is virtual, then

$$Ev(f(x)) = -i(Od(f(x)))^\sim, \tag{8.11}$$

$$Od(f(x)) = -i(Ev(f(x)))^\sim. \tag{8.12}$$

9. *If $f \in R(G)$ is covirtual, then (8.11) is also valid.*
   If $f \in R(G)$ is coactual, then (8.12) is also valid.

Proof. The following parts of the proof are numbered as the assertions of the theorem.

1. This assertion follows directly from definition of the Hilbert transform.

2. Proof of the first part of the statement follows from definition of the Hilbert transform since the function sign is real. The second part follows from Properties 5 and 3, respectively, since sign is an odd function.

3. $\overline{f^\sim} = -i\mathrm{sign}(\cdot)(\overline{f}) = \overline{i\mathrm{sign}(\cdot)f} = -(\overline{-i\mathrm{sign}(\cdot)f}) = -\overline{f^\sim}$.

4. Proof follows directly from definition of the Hilbert transform.

5. From definition 2 follows that $\mathrm{sign}^2(\cdot)$ is the identity coaxial function. It becomes apparent that since $\mathrm{sign}(x) = \mathrm{sign}^2(x) = 0$ for each $x \in SYM_V$, no inverse Hilbert transform exists for functions other than coaxial. Recall, however, that both actual and virtual functions are special kinds of coaxial functions.

6. Since f_a is axial, from property 1 it can be written as the sum of g_e and g_o. Then from property 6 we have

$$
\begin{aligned}
H(\mathbf{S}_{f_o}(w)) &= \mathbf{S}_{g_e}(w) = (\mathrm{sign}(x)g_o(x))^\top \\
&= (-i\mathrm{sign}(x)(ig_o(x)))^\top \\
&= ((ig_o(x))^\sim)^\top = (((i\mathbf{S}_{g_o}(w))^\perp)^\sim)^\top \\
&= (((isH(\mathbf{S}_{f_a}(w)))^\perp)^\sim)^\top,
\end{aligned}
$$

and similarly

$$
\begin{aligned}
sH(\mathbf{S}_{f_o}(w)) &= \mathbf{S}_{g_o}(w) = (g_o)(x))^\top \\
&= i(-i\mathrm{sign}(x)g_o(x))^\top = i((g_e(x))^\sim)^\top \\
&= i(((\mathbf{S}_{g_e}(w))^\perp)^\sim)^\top = i(((H(\mathbf{S}_{f_a}(w)))^\perp)^\sim)^\top.
\end{aligned}
$$

The proof is analogous in the case of virtual functions.

7. Let f be covirtual. Then f may be expressed as $f = f_a + g_{\mathrm{sym}}$, where f_a is actual and g_{sym} takes the same values as f for each $x \in SYM_V$ and is zero otherwise. It becomes apparent that for each $x \in G$, $\mathrm{sign}(x)g_{\mathrm{sym}}(x) = 0$, since g_{sym} is axial (and therefore also even). Particularly, $(g_{\mathrm{sym}})^{sym} = 0$.

It follows

$$
f = f_a + g_{\mathrm{sym}} = g_o + g_e + g_{\mathrm{sym}},
$$

where $Od(f) = g_o$ and $Ev(f) = g_e + g_{\mathrm{sym}}$.

From property 2 we have

$$
\begin{aligned}
sH(\mathbf{S}_f(w)) &= (Od(f))^\top = (g_o(x))^\top = i(-i\mathrm{sign}(x)g_e(x))^\top \\
&= i(-i\mathrm{sign}(x)g_e(x) - i\mathrm{sign}(x)g_{\mathrm{sym}}(x))^\top \\
&= i((g_e(x) + g_{\mathrm{sym}}(x))^\sim)^\top = i((Ev(f))^\sim)^\top \\
&= i(((H(\mathbf{S}_f(w)))^\perp)^\sim)^\top.
\end{aligned}
$$

Similarly for f a coactual function.

8. Since f_a is actual, then $f_a = g_e + g_o = Ev(f_a) + Od(f_a)$. From property 1 follows that $g_e(x) = \mathrm{sign}(x)g_o(x)$, hence:

$$
\begin{aligned}
Ev(f_a(x)) &= \mathrm{sign}(x)Od(f_a(x)) \\
&= (-i\mathrm{sign}(x))(iOd(f_a(x))) = (iOd(f_a)x)))^\sim
\end{aligned}
$$

$$
\begin{aligned}
Od(f_a(x)) &= \mathrm{sign}(x)Ev(f_a(x)) \\
&= (-i\mathrm{sign}(x))(iEv(f_a(x))) = (iEv(f_a(x)))^\sim.
\end{aligned}
$$

9. See the proof of assertion 7.

Consider the following example for the illustration of the assertions of the theorem.

Example 8.1 *Let G be the quaternion group Q_2 defined in Example 2.3. Note that the function f considered in this example is an actual function and, therefore, will be denoted here by f_a. In Table 8.1 we list the values of the sign_V function on G and illustrate the decomposition of the given f_a into an even function g_e and an odd function g_o by using assertion 1 of the Theorem 8.1. Their Fourier transforms $\mathbf{S}_{g_e}$ and $\mathbf{S}_{g_o}$ are given in Table 8.2. It is apparent that $\mathbf{S}_{g_e} = H(\mathbf{S}_{f_a})$ and $\mathbf{S}_{g_o} = sH(\mathbf{S}_{f_a})$ as is stated in assertion 6 of the Theorem 8.1.*

Table 8.1   The even and odd parts of the test function.

x	x'	$\text{sign}_V(x)$	$f_a(x)$	$g_e(x)$	$g_o(x)$
0	0	0	0	0	0
1	3	1	α	$\frac{\alpha}{2}$	$\frac{\alpha}{2}$
2	2	0	0	0	0
3	1	-1	0	$\frac{\alpha}{2}$	$-\frac{\alpha}{2}$
4	6	1	β	$\frac{\beta}{2}$	$\frac{\beta}{2}$
5	7	1	λ	$\frac{\lambda}{2}$	$\frac{\lambda}{2}$
6	4	-1	0	$\frac{\beta}{2}$	$-\frac{\beta}{2}$
7	5	-1	0	$\frac{\lambda}{2}$	$-\frac{\lambda}{2}$

$SYM_V = \{0, 2\}, \quad POS_V = \{1, 4, 5\}, \quad NEG_V = \{3, 6, 7\}$

Table 8.2   Fourier spectrum of the test function over C.

w	$8S_{g_e}(w)$	$8S_{g_o}(w)$
0	$\alpha + \beta + \lambda$	0
1	$-\alpha + \beta - \lambda$	0
2	$\alpha - \beta - \lambda$	0
3	$-\alpha - \beta + \lambda$	0
4	$\begin{bmatrix} 0 & 0 \\ 0 & 0 \end{bmatrix}$	$2\begin{bmatrix} -i\alpha & \beta + i\lambda \\ -\beta + i\lambda & i\alpha \end{bmatrix}$

8.3 HILBERT TRANSFORM IN FINITE FIELDS

In this section we will consider the definition of the Hilbert transform on finite non-Abelian groups for functions taking their values in a finite field admitting the existence of a Fourier transform.

Notice that some of the results from Section 8.2 have to be slightly modified so that they hold also in finite fields. However, some other no longer exists in this case. The main difference which should be appreciated is that the operation we denote by ∗ simply reduces to the transposition. In this setting the concepts of Hermitean and skew-Hermitean matrix should be reformulated.

Definition 8.5 *For the Fourier spectrum* $\mathbf{S}_f$ *of a function* $f \in P(G)$ *we define the field Hermitean part by* $fH(\mathbf{S}_f) = \mathbf{S}_{f_e}$, *and the field skew-Hermitean part by* $fsH(\mathbf{S}_f) = \mathbf{S}_{f_o}$, *where the functions* f_e *and* f_o *are defined in property 1.*

Definition 8.6 *The Hilbert transform* $f^\sim$ *of a function* $f \in P(G)$, *where* G *is a non-Abelian group, is defined under a given bijection* V *as the linear operator* $\sim$: $P(G) \to P(G)$ *given by:*

$$f^\sim(x) = sign(x)f(x), \quad \forall x \in G,$$

with $sign_V(\cdot)$ *as in Definition 8.2.*

It can be shown that except for Property 8.3, all other properties from Theorem 8.1 hold also in this case omitting simply the imaginary unit. Therefore, we have the following theorem.

Theorem 8.2 *The main properties of the Hilbert transform for functions belonging to* $P(G)$ *are*

1. *For each* $f, h \in P(G)$, *and* $a, b \in P$
   $(af + bh))^\sim(x) = af^\sim(x) + bh^\sim(x)$.

2. *If* $f \in P(G)$, *is even, then* $f^\sim$ *is odd, and if* f *is odd,* $f^\sim$ *is even.*

3. *For each* $f, h \in P(G), f^\sim(x)h(x) = f(x)h^\sim(x)$

4. *The inverse Hilbert transform of a coaxial function* $f \in P(G)$ *is given by* $f^\sim(f^\sim(x)) = -f(x)$. *Notice that the actual and virtual functions are also coaxial functions, and it follows*
   $f(x)h(x) = -f^\sim(x)h^\sim(x)$.

5. *Let* $f_a \in P(G)$ *be actual. Then the Hermitean and skew-Hermitean parts of its Fourier spectrum* $\mathbf{S}_f$ *are related by the Hilbert transform as shown below: For each* $w \in \{0, \dots, K-1\}$

$$fH(\mathbf{S}_{f_a}(w)) = (((fsH(\mathbf{S}_{f_a})))^\perp)^\sim)^\top, \tag{8.13}$$

$$f s H(\mathbf{S}_{f_a}(w)) = (((f H(\mathbf{S}_{f_a})))^{\perp})^{\sim})^{\top}, \tag{8.14}$$

Let $f \in P(G)$ be virtual. Then,

$$f H(\mathbf{S}_{f_v}(w)) = -(((f s H(\mathbf{S}_{f_v})))^{\perp})^{\sim})^{\top}, \tag{8.15}$$

$$f s H(\mathbf{S}_{f_v}(w)) = -(((f H(\mathbf{S}_{f_v})))^{\perp})^{\sim})^{\top}. \tag{8.16}$$

6. *Let $f \in P(G)$ be covirtual. Then equation (8.7) is also valid.*
   Let $f \in P(G)$ be coactual. Then equation (8.8) is also valid.

7. *Let $f \in P(G)$. Moreover, let $Ev(f)$ and $Od(f)$ denote the even and odd parts of f, respectively. These parts are related by the Hilbert transform as follows: If f is actual, then*

$$Ev(f(x)) = (Od(f(x)))^{\sim}, \tag{8.17}$$

$$Od(f(x)) = (Ev(f(x)))^{\sim}, \tag{8.18}$$

and if f is virtual, then

$$Ev(f(x)) = -(Od(f(x)))^{\sim}, \tag{8.19}$$

$$Od(f(x)) = -(Ev(f(x)))^{\sim}. \tag{8.20}$$

8. *If $f \in P(G)$ is covirtual, then (8.11) is also valid.*
   If $f \in P(G)$ is coactual, then (8.12) is also valid.

It follows from the properties stated in this theorem that the transform introduced by Definition 8.5 can be regarded as the Hilbert transform for functions on finite non-Abelian groups into finite fields representing a proper counterpart of the Hilbert transform introduced by Definition 8.3 and, further, as a counterpart of the classical Hilbert transform in L^2 as well as the Hilbert transform on finite Abelian groups, see for example, [1], [4].

Example 8.2 *Let G be the group described in Example 6.2. In Table 8.3 the sign function, an actual function and its even and odd parts are shown. In Table 8.4 the Fourier spectrum of this function and in Table 8.3 of its field Hermitean and field skew-Hermitean parts are given. We define the field Hermitean and field skew-Hermitean parts of the Fourier spectrum respectively by $f H(\mathbf{S}_{f_a} = \mathbf{S}_{f_{ae}}$, and $f s H(\mathbf{S}_{f_a} = \mathbf{S}_{f_{ao}}$, where $f_{ae}(x) = \frac{1}{2}(f(x) + f(x'))$ and $f_{ae}(x) = \frac{1}{2}(f(x) - f(x'))$.*
Note that $(\mathbf{S}_{f_a})^T = \mathbf{S}_{f_{ae}} - \mathbf{S}_{f_{ao}}$ and that formulas (8.15) and (8.16) are true.

Table 8.3 The even and odd parts of the test function.

x	x'	$\text{sign}_V(x)$	f_a	f_{ae}	f_{ao}
0	0	0	0	0	0
1	2	1	α	$\frac{\alpha}{2}$	$\frac{\alpha}{2}$
2	1	10	0	$\frac{\alpha}{2}$	$-\frac{\alpha}{2}$
3	3	0	0	0	0
4	4	0	0	0	0
5	5	0	0	0	0
6	6	0	0	0	0
7	8	1	β	$\frac{\beta}{2}$	$\frac{\beta}{2}$
8	7	10	0	$\frac{\beta}{2}$	$-\frac{\beta}{2}$
9	9	0	0	0	0
10	10	0	0	0	0
11	11	0	0	0	0

$POS_V = \{1, 7\}$,
$SYM_V = \{0, 3, 4, 5, 6, 9, 10, 11\}$,
$NEG_V = \{2, 8\}$

Table 8.4 Fourier spectrum of the test function.

w	$\mathbf{S}_{f_a}$
0	$\alpha + \beta$
1	$\alpha + \beta$
2	$\begin{bmatrix} 10\alpha + 10\beta & 6\alpha + 6\beta \\ 5\alpha + 5\beta & 10\alpha + 10\beta \end{bmatrix}$
3	$\alpha + 10\beta$
4	$\alpha + 10\beta$
5	$\begin{bmatrix} 10\alpha + \beta & 6\alpha + 5\beta \\ 5\alpha + 6\beta & 10\alpha + \beta \end{bmatrix}$

Table 8.5 Fourier spectrum of the test function in $GF(11)$.

w	$\mathbf{S}_{f_{ae}}$	$\mathbf{S}_{f_{ao}}$
0	$\alpha + \beta$	0
1	$\alpha + \beta$	0
2	$\begin{bmatrix} 10\alpha + 10\beta & 0 \\ 0 & 10\alpha + 10\beta \end{bmatrix}$	$\begin{bmatrix} 0 & 6\alpha + 6\beta \\ 5\alpha + 5\beta & 0 \end{bmatrix}$
3	$\alpha + 10\beta$	0
4	$\alpha + 10\beta$	0
5	$\begin{bmatrix} 10\alpha + \beta & 0 \\ 0 & 10\alpha + \beta \end{bmatrix}$	$\begin{bmatrix} 10\alpha + \beta & 6\alpha + 5\beta \\ 5\alpha + 6\beta & 0 \end{bmatrix}$

REFERENCES

1. Bljumin, S.L., Trakhtman, A.M., "Discrete Hilbert transform on a finite intervals", *Radiotehnika in Elektronika*, No. 7, 1977, 1390-1398 (in Russian).

2. Butzer, P.L., Nessel, R.J., *Fourier Analysis and Approximation*, Birkhauser Verlag, Basel and Stuttgart, 1971.

3. Čižek, V., "Discrete Hilbert transform", *IEEE Trans.*, Vol. AU-18, No. 4, 1970, 340-343.

4. Moraga, C., Salinas, L., "On Hilbert and Chrestenson transform on finite Abelian groups", *Proc. 16th Int. Symp. on Multiple-Valued Logic*, Blacksburg, VA, 1986.

5. Moraga, C., Salinas, L., "On Hilbert and Chrestenson transforms on finite Abelian groups", *Scientia*, 1988.

6. Moraga, C., "On the Hilbert and Zhang-Hartley transform in finite Abelian groups", *Journal of China Institute of Communications*, (8), 3, 1987, 10-20.

7. Moraga, C., Stanković, R.S., "Hilbert transform in finite fields", *Proc. XXXVII Conf. for ETAN*, 20.-23.9.1993, Beograd, 59-64.

8. Stanković, R.S., Moraga, C., "Hilbert transform on finite non-Abelian groups", in Y. Baozong, Ed., *Proc. Int. Conf. on Signal Processing/Beijing '90*, International Academic Publishers, Beijing, China, 1990, 1179-1182.

9. Stanković, R.S., Stojić, M.R., Bogdanović, S.M., *Fourier Representation of Signals*, Naučna knjiga, Beograd, 1988, in Serbian.

10. Vlasenko, V.A., Shodin, O.I., *Microprocessors Systems for Quality Control of Digital Devices*, Tehnika, Kiev, 1990 (in Russian).

Index

Printed in the United States
By Bookmasters